SpringerBriefs in Applied Sciences and Technology

# PoliMI SpringerBriefs

Springer, in cooperation with Politecnico di Milano, publishes the PoliMI Springer-Briefs, concise summaries of cutting-edge research and practical applications across a wide spectrum of fields. Featuring compact volumes of 50 to 125 (150 as a maximum) pages, the series covers a range of contents from professional to academic in the following research areas carried out at Politecnico:

- Aerospace Engineering
- Bioengineering
- Electrical Engineering
- Energy and Nuclear Science and Technology
- Environmental and Infrastructure Engineering
- Industrial Chemistry and Chemical Engineering
- Information Technology
- Management, Economics and Industrial Engineering
- Materials Engineering
- Mathematical Models and Methods in Engineering
- Mechanical Engineering
- Structural Seismic and Geotechnical Engineering
- Built Environment and Construction Engineering
- Physics
- Design and Technologies
- Urban Planning, Design, and Policy

Matteo Pozzi • Francesco Braghin

# Structural Optimization for Nonlinear Dynamics

## Towards the Optimization of Dynamic Nonlinearities in MEMS Devices

Matteo Pozzi
Dipartimento di Meccanica
Politecnico di Milano
Milano, Italy

Francesco Braghin
Dipartimento di Meccanica
Politecnico di Milano
Milano, Italy

ISSN 2191-530X ISSN 2191-5318 (electronic)
SpringerBriefs in Applied Sciences and Technology
ISSN 2282-2577 ISSN 2282-2585 (electronic)
PoliMI SpringerBriefs
ISBN 978-3-032-22599-3 ISBN 978-3-032-22600-6 (eBook)
https://doi.org/10.1007/978-3-032-22600-6



This Springer imprint is published by the registered company Springer Nature Switzerland AG
The registered company address is: Gewerbestrasse 11, 6330 Cham, Switzerland

*To our families and friends*

# Preface

The increasing complexity of modern mechanical systems demands advanced methods for predicting and optimizing their nonlinear dynamic behavior. Classical linear approaches often prove inadequate in realistic operating regimes, where phenomena such as amplitude-dependent frequency shifts, internal resonances, and modal interactions play a decisive role. These challenges are particularly relevant in systems operating under complex, time-varying conditions, including microelectromechanical systems (MEMS), robotics, and advanced structural components.

This book addresses these challenges by presenting a rigorous yet accessible framework that integrates nonlinear dynamics, model reduction, and structural optimization. Central to this framework is Spectral Submanifold (SSM) theory, a powerful mathematical tool for the systematic analysis of nonlinear dynamical systems. By combining SSM-based reduced-order modeling with advanced optimization techniques, the book introduces an SSM-based optimization methodology for tailoring and enhancing the nonlinear dynamic response of mechanical systems.

The book begins by introducing the motivating challenges and objectives, followed by a review of the essential theoretical foundations and computational tools, including SSM theory as well as classical parametric and topology optimization techniques. It then presents the proposed framework in detail, supported by representative numerical results and applications, and concludes with a discussion of broader implications, limitations, and directions for future research.

Throughout the text, we aim to offer a balance between theoretical rigor and practical relevance, providing the reader with both a solid understanding of the underlying principles and the tools needed to apply them in real-world scenarios.

This book is intended for graduate students, researchers, and practicing engineers in applied mechanics, nonlinear dynamics, and structural optimization who seek a rigorous yet practical understanding of this emerging area.

Milan, Italy
October 2025

Matteo Pozzi
Francesco Braghin

**Acknowledgements** This work builds upon my doctoral research, carried out over three years, during which I had the privilege of meeting and collaborating with many professors, researchers, and fellow doctoral students. Among them, I wish to express my deepest gratitude to Prof. Francesco Braghin, who supervised and co-authored this work. A special mention also goes to Dr. Giacomo Bonaccorsi, who first introduced me to the field of structural and topology optimization, and to Dr. Jacopo Marconi, for his invaluable guidance in nonlinear dynamics.

Finally, I would like to extend my heartfelt thanks to my family, Lucrezia, and my lifelong friends for their constant love and support throughout this journey. This work is dedicated to them.

Matteo Pozzi

**Competing Interests** The authors have no competing interests to declare that are relevant to the content of this manuscript.

# Contents

# Chapter 1
# Introduction

**Abstract** This chapter introduces the motivation, objectives, and structure of the book. It discusses the challenges of optimizing the nonlinear dynamic response of mechanical and microelectromechanical systems (MEMS). The chapter outlines how Spectral Submanifold (SSM) theory can be leveraged to bridge the gap between nonlinear dynamics and modern optimization techniques, setting the stage for the developments presented in the following parts.

Micro Electro Mechanical Systems (MEMS) are devices that integrate mechanical moving parts with electronic components on a single chip. Despite some non-planar features, MEMS are primarily two-dimensional and are batch fabricated in a significant number of identical components on a single silicon wafer (Fig. 1.1). MEMS technology evolved from semiconductor fabrication technology, and its structure geometry is generated through photolithography and etching. These devices have revolutionized the fields of sensing through their miniaturized, cost-effective, and energy-efficient designs. Among the various classes of MEMS devices, MEMS inertial sensors [17], including accelerometers and gyroscopes, are fundamental in enabling motion tracking and orientation sensing across a wide array of applications, from consumer electronics (e.g., smartphones, gaming consoles) to automotive systems, industrial robotics, aerospace navigation, and biomedical instrumentation.

MEMS gyroscopes typically operate based on the Coriolis effect, where a vibrating mechanical structure experiences an apparent force proportional to the angular rate of rotation [1]. The basic architecture of a vibratory gyroscope consists of a proof mass suspended above the substrate by flexible beams, commonly referred to as springs. In a simple configuration, the proof mass is free to oscillate along two main orthogonal directions, the drive and sense modes, each associated with its natural frequency (Fig. 1.2).

Under typical operating conditions, the proof mass is driven into resonance along the drive direction, usually through electrostatic actuation. When an angular rate $\Omega_{ext}$ is applied, the Coriolis effect generates an apparent force that causes the proof mass to move along the sense direction. This induced motion is proportional to the

 
M. Pozzi, F. Braghin, *Structural Optimization for Nonlinear Dynamics*,
PoliMI SpringerBriefs, https://doi.org/10.1007/978-3-032-22600-6_1

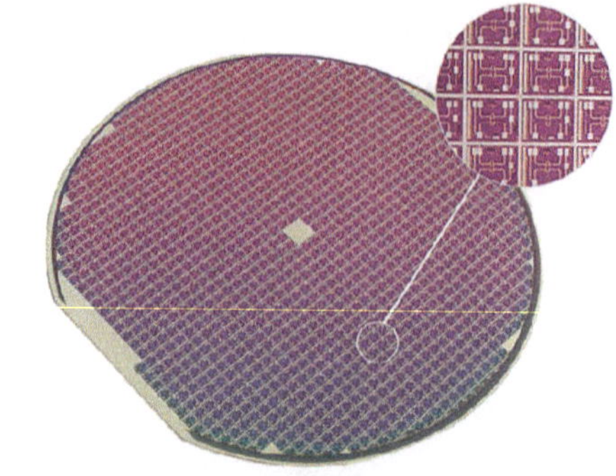

Fig. 1.1 Silicon wafer with fabricated MEMS devices

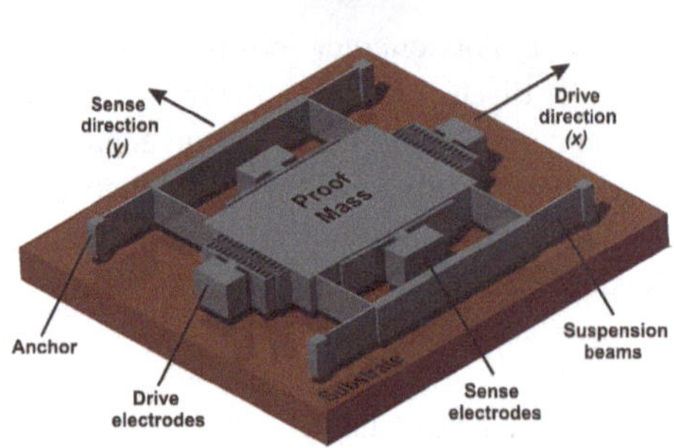


Fig. 1.2 Scheme of a single axis MEMS gyroscope [1]

angular velocity and can be measured, typically using capacitive sensing electrodes, to infer the input rotation.

The drive resonant frequency $\omega_d$ is typically set in the range of 20–30 kHz to mitigate acoustic and mechanical issues associated with vibrating masses [1]. To maximize the Coriolis response, it is generally advantageous to operate the gyroscope at or near the peak of the sense-mode resonance. This condition is commonly achieved by matching the sense resonant frequency, $\omega_s$, with the drive frequency, $\omega_d$, which significantly amplifies the mechanical response of the sense mode to angular rate inputs. However, such frequency matching makes the system highly sensitive to parameter variations that may shift the resonant frequencies or affect damping characteristics. For example, the sense mode amplification of a system with $\omega_s = 10\,\text{kHz}$ and Quality factor $Q_s = 1000$ drops by over 55% for a 10 Hz relative shift between sense and drive modes [1]. For this reason, it is often preferable to introduce a controlled frequency mismatch between the sense and drive modes to enhance robustness:

$$\Delta\omega = |\omega_d - \omega_s|, \tag{1.1}$$

where typical mismatch values are around 2–10% of $\omega_s$.

The sensitivity, bandwidth, and long-term stability of a MEMS gyroscope depend critically on the design of its vibrating structure, also referred to as a MEMS resonator, and how it couples mechanical energy between its drive and sense modes. As MEMS devices are increasingly required to operate under higher actuation levels, broader bandwidths, and more demanding environmental conditions, nonlinear effects in their dynamic response become increasingly significant and must be carefully considered in both design and analysis. In MEMS gyroscopes, these nonlinearities arise from various sources, including mechanical, electrostatic, and environmental factors [17, 24, 37]. Mechanical nonlinearities, such as those induced by large deformations and mode coupling, often dominate the system's response, particularly at high amplitudes. Electrostatic actuation and sensing introduce additional nonlinearities due to their inherent voltage and displacement dependencies, which can lead to phenomena such as pull-in instability and electrostatic spring softening. Moreover, other effects such as temperature fluctuations and residual stresses from fabrication further contribute to variability and performance degradation. Among these effects, geometric nonlinearities are of particular interest, as they directly depend on the shape and layout of the device.

The increasing number of applications utilizing nonlinear phenomena emphasizes the necessity of considering nonlinear effects early in the design process. In many cases, nonlinearities are no longer seen as undesirable byproducts but are intentionally exploited to enhance system performance. Notable examples include frequency division [36], vibration mitigation [3], nonlinear energy sinks and targeted energy transfer [44], micro- and nano-electromechanical resonators [2, 26, 32], and MEMS sensors [6, 29, 45].

The growing complexity of such systems has driven the development of advanced analytical and computational Reduced Order Model (ROM) techniques [42, 43] for studying nonlinear dynamic behavior. A common strategy involves computing the frequency-amplitude relation of Nonlinear Normal Modes (NNMs) [18, 40] via *backbone curves* [5]. These curves are typically obtained using numerical continuation methods, such as the shooting method [30], the collocation method [9], or the Harmonic Balance Method (HBM) [20]. Alternatively, backbone curves can be derived from the reduced dynamics on Spectral Submanifolds (SSMs) [14], an approach that has proven highly effective for high-dimensional mechanical systems [15, 41].

The range of available numerical methods, together with the increasing interest in nonlinear applications, provides an excellent playground for developing optimization techniques aimed at tailoring the nonlinear dynamic response of mechanical systems. Approaches based on NNMs and HBM have been applied to control hardening behavior in resonators and frame structures [11–13, 23]. Similarly, shape optimization techniques have been employed to tune eigenfrequencies and modal coupling coefficients in geometrically nonlinear MEMS gyroscopes [31, 39]. Other methods include nonlinear synthesis for backbone tailoring [10] and gradient-free optimization [27]. In topology optimization [4], the Equivalent Static Load method [7] has been used for nonlinear structures [19, 21, 28], while [8, 25] addressed eigenvalue optimization with frequency-dependent materials.

Despite these advances, designing structures with tailored nonlinear responses remains challenging. High-dimensional parameter spaces, the sensitivity of numerical solvers to user-defined settings, and the evolving nature of the system during optimization all contribute to computational complexity. For instance, methods such as HBM and continuation require careful tuning of harmonics, step size, and arc-length parameters to ensure convergence. Moreover, limitations in sensitivity computation can affect the reliability of optimization outcomes [38].

## 1.1 Spectral Submanifold-Based Optimization

The framework presented in this book addresses these challenges by introducing an optimization methodology based on Spectral Submanifolds (SSMs). The key advantage of SSM reduction (Fig. 1.3) lies in its ability to yield analytical expressions for backbone curves, which are then used as optimization targets. This eliminates the dependency on user-defined numerical parameters inherent in methods such as HBM or collocation. The only adjustable quantity is the expansion order of the SSM, which can be automatically tuned based on the residual of the invariance equation to maintain a prescribed reduction accuracy.

The initial implementation focused on conservative systems, where the SSM formulation reduces to the Lyapunov Subcenter Manifold (LS$\mathcal{M}$[1]) [16], from which the conservative backbone curve is derived [33]. Sensitivities were initially computed via direct differentiation, a process that becomes computationally expensive as the number of design variables increases. Moreover, the LS$\mathcal{M}$ formulation employed a tensor-based notation [15], which, while rigorous, was not the most efficient for large-scale polynomial expansions.

To improve computational performance, subsequent work adopted a multi-index notation [41], enabling faster and more scalable computations. This was demonstrated in [35], where topology optimization was applied to tune the hardening/softening behavior of conservative nonlinear systems. The adjoint sensitivity formulation derived in [22], adapted here to the undamped case, allowed opti-

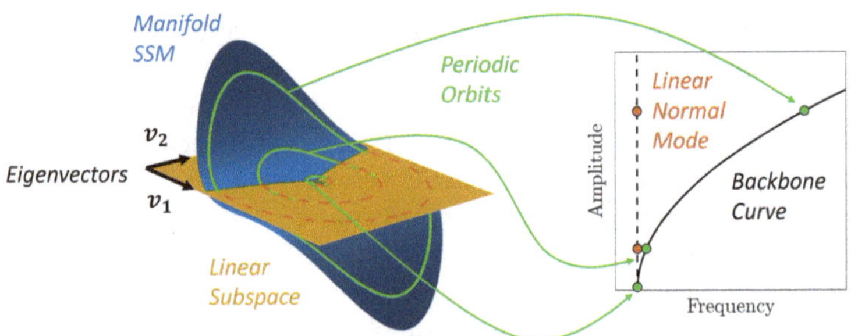


**Fig. 1.3** Example of SSM reduction

[1] The acronym LS$\mathcal{M}$ is used here to distinguish the Lyapunov Subcenter Manifold from the Level Set Method, commonly used in topology optimization literature.

mization with over 20,000 design variables and approximately 40,000 degrees of freedom.

However, this explicit formulation was limited to cubic-order LS$\mathcal{M}$ computations in conservative settings and optimized the backbone curve in the reduced space rather than the physical space. Since the shape of backbone curves depends on the chosen parametrization [15], optimizing in reduced coordinates can yield discrepancies when mapped to the physical response.

To overcome these limitations, the framework is generalized to lightly damped systems and extended to arbitrary polynomial orders [34]. Adjoint sensitivity expressions are derived for both the SSM and the physical-space backbone curve, enabling direct optimization of the frequency-amplitude relation in physical coordinates. The multi-index representation ensures computational efficiency, while an adaptive expansion strategy maintains control over model accuracy during optimization.

## 1.2 Outline of the Book

This book is structured as follows. Chapter 1 introduces the context, motivation, and objectives of the study, outlining the theoretical foundations and the development of the SSM-based optimization framework.

The following chapters review the state-of-the-art in nonlinear dynamics, model reduction, and structural optimization. In particular, Chap. 2 discusses SSM theory for analyzing the nonlinear dynamic response of mechanical systems. Chapter 3 presents the principles of structural optimization, with emphasis on size optimization for MEMS design. Chapter 4 introduces topology optimization and its application to linear dynamic systems.

Following this, the core of the book details the proposed SSM-based optimization framework for tailoring the nonlinear dynamic response of mechanical systems. Chapter 5 defines the optimization objectives and workflow. Chapter 6 derives analytical sensitivity expressions using both direct and adjoint methods. Chapter 7 presents and discusses numerical applications in parametric and topology optimization, demonstrating the method's accuracy and efficiency.

Finally, Chap. 8 summarizes the main findings and discusses future research directions.

## References

1. Acar C, Shkel A (2009) MEMS vibratory gyroscopes. Springer, Berlin (2009). https://doi.org/10.1007/978-0-387-09536-3
2. Asadi K, Yeom J, Cho H (2021) Strong internal resonance in a nonlinear, asymmetric microbeam resonator. Microsyst Nanoeng 7:9. https://doi.org/10.1038/s41378-020-00230-1

3. Bellet R, Cochelin B, Herzog P, Mattei PO (2010) Experimental study of targeted energy transfer from an acoustic system to a nonlinear membrane absorber. J Sound Vib 329:2768–2791. https://doi.org/10.1016/j.jsv.2010.01.029
4. Bendsøe MP, Sigmund O (2004) Topology optimization: theory, methods, and applications. Springer, Berlin. https://doi.org/10.1007/978-3-662-05086-6
5. Breunung T, Haller G (2018) Explicit backbone curves from spectral submanifolds of forced-damped nonlinear mechanical systems. Proc R Soc A Math Phys Eng Sci 474:20180083. https://doi.org/10.1098/rspa.2018.0083
6. Chen C, Zanette DH, Czaplewski DA, Shaw S, Løpez D (2017) Direct observation of coherent energy transfer in nonlinear micromechanical oscillators. Nat Commun 8:15523. https://doi.org/10.1038/ncomms15523
7. Choi W, Park G (2002) Structural optimization using equivalent static loads at all time intervals. Comput Methods Appl Mech Eng 191:2105–2122. https://doi.org/10.1016/S0045-7825(01)00373-5
8. Dalklint A, Wallin M, Tortorelli DA (2022) Eigenfrequency constrained topology optimization of finite strain hyperelastic structures. Struct Multidiscip Optim 61:2577–2594. https://doi.org/10.1007/s00158-020-02557-9
9. Dankowicz H, Schilder F (2013) Recipes for continuation. Society for Industrial and Applied Mathematics, Philadelphia. https://doi.org/10.1137/1.9781611972573
10. Detroux T, Noël JP, Kerschen G (2021) Tailoring the resonances of nonlinear mechanical systems. Nonlinear Dyn. 103:3611–3624. https://doi.org/10.1007/s11071-020-06002-w
11. Dou S, Jensen JS (2015) Optimization of nonlinear structural resonance using the incremental harmonic balance method. J. Sound Vib. 334:239–254. https://doi.org/10.1016/j.jsv.2014.08.023
12. Dou S, Jensen JS (2016) Optimization of hardening/softening behavior of plane frame structures using nonlinear normal modes. Comput. Struct. 164:63–74. https://doi.org/10.1016/j.compstruc.2015.11.001
13. Dou S, Strachan BS, Shaw SW, Jensen JS (2015) Structural optimization for nonlinear dynamic response. Philos. Trans. R. Soc. A Math. Phys. Eng. Sci. 373:20140408. https://doi.org/10.1098/rsta.2014.0408
14. Haller G, Ponsioen S (2016) Nonlinear normal modes and spectral submanifolds: existence, uniqueness and use in model reduction. Nonlinear Dyn. 86:1493–1534. https://doi.org/10.1007/s11071-016-2974-z
15. Jain S, Haller G (2022) How to compute invariant manifolds and their reduced dynamics in high-dimensional finite element models. Nonlinear Dyn. 107:1417–1450. https://doi.org/10.1007/s11071-021-06957-4
16. Kelley A (1969) Analytic two-dimensional subcenter manifolds for systems with an integral. Pac. J. Math. 29:335–350. https://doi.org/10.2140/pjm.1969.29.335
17. Kempe V (2011) Inertial MEMS. Cambridge University Press, Cambridge. https://doi.org/10.1017/CBO9780511933899
18. Kerschen G, Peeters M, Golinval J, Vakakis A (2009) Nonlinear normal modes, part i: a useful framework for the structural dynamicist. Mech. Syst. Signal Process. 23:170–194. https://doi.org/10.1016/j.ymssp.2008.04.002
19. Kim YI, Park GJ (2010) Nonlinear dynamic response structural optimization using equivalent static loads. Comput. Methods Appl. Mech. Eng. 199:660–676. https://doi.org/10.1016/j.cma.2009.10.014
20. Krack M, Gross J (2019) Harmonic balance for nonlinear vibration problems. Springer, Berlin. https://doi.org/10.1007/978-3-030-14023-6
21. Lee HA, Park GJ (2015) Nonlinear dynamic response topology optimization using the equivalent static loads method. Comput Methods Appl Mech Eng 283:956–970. https://doi.org/10.1016/j.cma.2014.10.015
22. Li M (2024) Explicit sensitivity analysis of spectral submanifolds of mechanical systems. Nonlinear Dyn 112:16733–16770. https://doi.org/10.1007/s11071-024-09947-4

23. Li LL, Polunin PM, Dou S, Shoshani O, Strachan BS, Jensen JS, Shaw SW, Turner KL (2017) Tailoring the nonlinear response of mems resonators using shape optimization. Appl Phys Lett 110:081902. https://doi.org/10.1063/1.4976749
24. Li D, Shaw SW, Polunin PM (2022) Computational modeling of nonlinear dynamics and its utility in mems gyroscopes. J Struct Dyn. 1:217–235. https://doi.org/10.25518/2684-6500.96
25. Li Q, Wu Q, Dou S, Wang J, Liu S, Chen W (2022) Nonlinear eigenvalue topology optimization for structures with frequency-dependent material properties. Mech Syst Signal Process 170:108835. https://doi.org/10.1016/j.ymssp.2022.108835
26. Li Z, Xu M, Norte RA, Aragón AM, Steeneken PG, Alijani F (2024) Strain engineering of nonlinear nanoresonators from hardening to softening. Commun Phys 7:53. https://doi.org/10.1038/s42005-024-01543-7
27. Li Z, Alijani F, Sarafraz A, Xu M, Norte RA, Aragón AM, Steeneken PG (2025) Finite element-based nonlinear dynamic optimization of nanomechanical resonators. Microsyst Nanoeng 11:16. https://doi.org/10.1038/s41378-024-00854-7
28. Lu S, Zhang Z, Guo H, Park GJ, Zuo W (2021) Nonlinear dynamic topology optimization with explicit and smooth geometric outline via moving morphable components method. Struct Multidisciplinary Optim 64:2465–2487. https://doi.org/10.1007/s00158-021-03000-3
29. Marconi J, Bonaccorsi G, Giannini D, Falorni L, Braghin F (2021) Exploiting nonlinearities for frequency-matched mems gyroscopes tuning. In: 2021 IEEE international symposium on inertial sensors and systems (INERTIAL). IEEE, Piscataway, pp 1–4. https://doi.org/10.1109/INERTIAL51137.2021.9430478
30. Peeters M, Viguié R, Sérandour G, Kerschen G, Golinval JC (2009) Nonlinear normal modes, part ii: toward a practical computation using numerical continuation techniques. Mech Syst Signal Process 23:195–216. https://doi.org/10.1016/j.ymssp.2008.04.003
31. Polunin PM, Shaw SW (2022) Maximizing the rate sensitivity of resonating gyroscopes using nonlinear shape optimization. J Micromech Microeng 32:064003. https://doi.org/10.1088/1361-6439/ac6c74
32. Polunin PM, Yang Y, Dykman MI, Kenny TW, Shaw SW (2016) Characterization of mems resonator nonlinearities using the ringdown response. J Microelectromech Syst 25:297–303. https://doi.org/10.1109/JMEMS.2016.2529296
33. Pozzi M, Marconi J, Jain S, Braghin F (2024) Backbone curve tailoring via lyapunov subcenter manifold optimization. Nonlinear Dyn 112:15719–15739. https://doi.org/10.1007/s11071-024-09881-5
34. Pozzi M, Marconi J, Jain S, Li M, Braghin F (2025) Adjoint sensitivities for the optimization of nonlinear structural dynamics via spectral submanifolds. Proc R Soc A Math Phys Eng Sci 481:20250244. https://doi.org/10.1098/rspa.2025.0244
35. Pozzi M, Marconi J, Jain S, Li M, Braghin F (2025) Topology optimization of nonlinear structural dynamics with invariant manifold-based reduced order models. Struct Multidiscip Optim 68:72. https://doi.org/10.1007/s00158-025-04010-1
36. Qalandar KR, Strachan BS, Gibson B, Sharma M, Ma A, Shaw SW, Turner KL (2014) Frequency division using a micromechanical resonance cascade. Appl Phys Lett 105:244103. https://doi.org/10.1063/1.4904465
37. Rhoads JF, Shaw SW, Turner KL (2010) Nonlinear dynamics and its applications in micro- and nanoresonators. J Dyn Syst Measur Control 132:034001. https://doi.org/10.1115/1.4001333
38. Saccani A, Marconi J, Tiso P (2023) Sensitivity analysis of nonlinear frequency response of defected structures. Nonlinear Dyn 111:4027–4051. https://doi.org/10.1007/s11071-022-08064-4
39. Schiwietz D, Hörsting M, Weig EM, Wenzel M, Degenfeld-Schonburg P (2025) Shape optimization of geometrically nonlinear modal coupling coefficients: an application to mems gyroscopes. Sci Rep 15:10957. https://doi.org/10.1038/s41598-025-95412-0
40. Shaw S, Pierre C (1993) Normal modes for non-linear vibratory systems. J Sound Vib 164:85–124. https://doi.org/10.1006/jsvi.1993.1198

41. Thurnher T, Haller G, Jain S (2024) Nonautonomous spectral submanifolds for model reduction of nonlinear mechanical systems under parametric resonance. Chaos Interdiscip J Nonlinear Sci 34:73127. https://doi.org/10.1063/5.0168431
42. Tiso P, Mahdiabadi MK, Marconi J (2021) Modal methods for reduced order modeling, pp. 97–138. De Gruyter, Beijing (2021). https://doi.org/10.1515/9783110498967-004
43. Touzé C, Vizzaccaro A, Thomas O (2021) Model order reduction methods for geometrically nonlinear structures: a review of nonlinear techniques. Nonlinear Dyn 105:1141–1190. https://doi.org/10.1007/s11071-021-06693-9
44. Vakakis AF, Gendelman OV, Bergman LA, Mojahed A, Gzal M (2022) Nonlinear targeted energy transfer: state of the art and new perspectives. Nonlinear Dyn 108:711–741. https://doi.org/10.1007/s11071-022-07216-w
45. Zhang W, Baskaran R, Turner KL (2002) Effect of cubic nonlinearity on auto-parametrically amplified resonant mems mass sensor. Sens Actuat A Phys 102:139–150. https://doi.org/10.1016/S0924-4247(02)00299-6

# Chapter 2
# Spectral Submanifold Theory

**Abstract** This chapter outlines the key concepts of Spectral Submanifold (SSM) theory, used to analyze the nonlinear dynamic response of mechanical systems. It describes the procedure to compute the autonomous and non-autonomous coefficients of a two-dimensional manifold and details the derivation of frequency response and backbone curves.

## 2.1 Introduction

In the context of nonlinear dynamics, the Spectral Submanifold (SSM) theory has proven to be a robust framework for analyzing nonlinear systems. It has been successfully applied to a range of engineering problems, specifically for the computation of forced response curves, backbone curves, and detached resonance curves (isolas) via exact reduced-order models in high-dimensional finite element models [8, 9].

For a comprehensive review of the general SSM theory, including all the mathematical details behind it, the reader is referred to [2–4, 12, 15]. In addition, the `SSMTool` [5] provides a MATLAB-based implementation for the computation of SSM and forced response curves in generic finite element models.

In this work, we focus on second-order, lightly damped, geometrically nonlinear mechanical systems subject to a single-harmonic external periodic excitation. To derive analytical expressions for the backbone and frequency response curves, we employ the two-dimensional SSM associated with a single master mode. The remainder of this chapter outlines the SSM-reduction framework under these assumptions.

 
M. Pozzi, F. Braghin, *Structural Optimization for Nonlinear Dynamics*,
PoliMI SpringerBriefs, https://doi.org/10.1007/978-3-032-22600-6_2

## 2.2 System Description

Consider a lightly damped, geometrically nonlinear mechanical system, described by the following second-order equation of motion:

$$\mathbf{M}\ddot{\boldsymbol{x}} + \mathbf{C}\dot{\mathbf{x}} + \mathbf{K}\boldsymbol{x} + \boldsymbol{f}(\boldsymbol{x}) = \varepsilon \boldsymbol{g} \cos(\Omega t), \tag{2.1}$$

where $\boldsymbol{x} \in \mathbb{R}^n$ represents the displacement vector with $n$ degrees of freedom. The matrices $\mathbf{M}$, $\mathbf{C}$, and $\mathbf{K}$ are the mass, damping, and stiffness matrices, respectively, each belonging to $\mathbb{R}^{n\times n}$. The vector $\boldsymbol{g}$ denotes the periodic external force with amplitude $\varepsilon$ and frequency $\Omega$, while $\boldsymbol{f} \in \mathbb{R}^n$ is the vector of displacement-dependent nonlinear internal forces.

In this work, the nonlinear internal force $\boldsymbol{f}$ is given as a linear combination of quadratic and cubic terms [10]:

$$\boldsymbol{f}(\boldsymbol{x}) = \boldsymbol{f}_2(\boldsymbol{x}, \boldsymbol{x}) + \boldsymbol{f}_3(\boldsymbol{x}, \boldsymbol{x}, \boldsymbol{x}), \tag{2.2}$$

where $\boldsymbol{f}_2$ and $\boldsymbol{f}_3$ denote the second- and third-order nonlinearities, respectively.

These are efficiently managed using the tensor notation [10] as

$$f^i = T_2^{ijk} x^j x^k + T_3^{ijkl} x^j x^k x^l, \tag{2.3}$$

where $f^i$ is the $i$-th component of $\boldsymbol{f}$, and the tensors $T_2 \in \mathbb{R}^{n\times n\times n}$ and $T_3 \in \mathbb{R}^{n\times n\times n\times n}$ collect the coefficients of the quadratic and cubic nonlinear terms. Here, superscripts are used to indicate tensor and vector components (not exponents).

The damping matrix $\mathbf{C}$ is defined according to the Rayleigh damping model:

$$\mathbf{C} = \alpha_R \mathbf{M} + \beta_R \mathbf{K}, \tag{2.4}$$

where $\alpha_R$ and $\beta_R$ are the mass and stiffness coefficients, respectively.

The second-order system (2.1) can be reformulated into its first-order form [4]:

$$\mathbf{B}\dot{z} = \mathbf{A}z + \boldsymbol{F}(z) + \varepsilon \boldsymbol{G}(\Omega t), \tag{2.5}$$

where

$$\begin{aligned} z &= \begin{bmatrix} \boldsymbol{x} \\ \dot{\boldsymbol{x}} \end{bmatrix}, \quad \mathbf{B} = \begin{bmatrix} \mathbf{C} & \mathbf{M} \\ \mathbf{M} & \mathbf{0} \end{bmatrix}, \quad \mathbf{A} = \begin{bmatrix} -\mathbf{K} & \mathbf{0} \\ \mathbf{0} & \mathbf{M} \end{bmatrix}, \\ \boldsymbol{F}(z) &= \begin{bmatrix} -\boldsymbol{f}(\boldsymbol{x}) \\ \mathbf{0} \end{bmatrix}, \quad \boldsymbol{G}(\Omega t) = \begin{bmatrix} \boldsymbol{g}\cos(\Omega t) \\ \mathbf{0} \end{bmatrix}. \end{aligned} \tag{2.6}$$

The eigenvalues $\lambda_i \in \mathbb{C}$ are determined by solving the generalized right and left eigenvalue problems [15]:

$$(\mathbf{A} - \lambda_i \mathbf{B})\, \boldsymbol{v}_i = \mathbf{0}, \quad \boldsymbol{u}_i^* (\mathbf{A} - \lambda_i \mathbf{B}) = \mathbf{0}, \tag{2.7}$$

where $(\bullet)^*$ denotes the conjugate transpose, and $\boldsymbol{v}_i, \boldsymbol{u}_i \in \mathbb{C}^{2n}$ are the associated right and left eigenvectors.

The eigenstructure of the first-order system can be related to the eigen-properties of the corresponding undamped second-order system [15] through:

$$\lambda_1 = \bar{\lambda}_2 = \lambda = -\xi\omega + \mathrm{i}\omega\sqrt{1-\xi^2} = \alpha + \mathrm{i}\omega_d, \tag{2.8}$$

$$\boldsymbol{v}_1 = \bar{\boldsymbol{v}}_2 = \begin{bmatrix} \boldsymbol{\varphi} \\ \lambda\boldsymbol{\varphi} \end{bmatrix}, \quad \boldsymbol{u}_1 = \bar{\boldsymbol{u}}_2 = \begin{bmatrix} \boldsymbol{\varphi} \\ \bar{\lambda}\boldsymbol{\varphi} \end{bmatrix}, \tag{2.9}$$

where $(\bar{\bullet})$ indicates complex conjugation, and $\boldsymbol{\varphi} \in \mathbb{R}^n$ and $\omega \in \mathbb{R}$ are solutions of the undamped eigenvalue problem

$$\left(\mathbf{K} - \omega^2\mathbf{M}\right)\boldsymbol{\varphi} = \mathbf{0}. \tag{2.10}$$

The mode shapes $\boldsymbol{\varphi}$ are mass-normalized according to

$$\boldsymbol{\varphi}^T\mathbf{M}\boldsymbol{\varphi} = 1, \quad \boldsymbol{\varphi}^T\mathbf{K}\boldsymbol{\varphi} = \omega^2. \tag{2.11}$$

Using the Rayleigh damping expression (2.4) and the normalization condition (2.11), the modal damping $\delta$ is given by

$$\delta = \boldsymbol{\varphi}^T\mathbf{C}\boldsymbol{\varphi} = \alpha_R + \beta_R\omega^2. \tag{2.12}$$

The damping ratio $\xi$ then follows as

$$\xi = \frac{\delta}{2\omega} = \frac{\alpha_R + \beta_R\omega^2}{2\omega}. \tag{2.13}$$

Finally, the assumption of light damping corresponds to the condition $\xi \ll 1$. Using the damping ratio, the Q-factor is defined as

$$Q = \frac{1}{2\xi} = \frac{\omega}{\alpha_R + \beta_R\omega^2}. \tag{2.14}$$

**Damping**
The coefficients $\alpha_R$ and $\beta_R$ may either be held constant during the optimization or selected to achieve a target Q-factor. In the latter case, two easy choices are available:

$$\alpha_R = \frac{\omega}{Q} \quad \text{and} \quad \beta_R = 0,$$

$$\alpha_R = 0 \quad \text{and} \quad \beta_R = \frac{1}{\omega Q}.$$

## 2.3 Spectral Submanifold Reduction

We provide here the basic steps used to compute the two-dimensional SSM of the system (2.1).

Call $E$ the two-dimensional master (linear) subspace [4] spanned by the eigenvectors $\boldsymbol{v}_1$ and $\boldsymbol{v}_2$ associated with the master eigenfrequency $\omega$ and mode shape $\boldsymbol{\varphi}$.

Let $\boldsymbol{R}_\varepsilon : \mathbb{C}^2 \times \mathbb{S} \to \mathbb{C}^2$ be a two-dimensional parametrization of the reduced dynamics, such that

$$\dot{\boldsymbol{p}} = \boldsymbol{R}_\varepsilon(\boldsymbol{p}, \phi) = \boldsymbol{R}(\boldsymbol{p}) + \varepsilon \boldsymbol{S}(\boldsymbol{p}, \phi), \tag{2.15}$$

where $\boldsymbol{p} \in \mathbb{C}^2$ is the vector of reduced coordinates and $\phi \in \mathbb{S}$ is the phase of the periodic excitation term belonging to the circle[1] $\mathbb{S}$, such that $\dot{\phi} = \Omega$. The terms $\boldsymbol{R}$ and $\boldsymbol{S}$ are the autonomous and non-autonomous parametrizations, respectively. Using the coordinates $\boldsymbol{p}$, let $\boldsymbol{W}_\varepsilon : \mathbb{C}^2 \times \mathbb{S} \to \mathbb{R}^{2n}$ be a parametrization of the SSM attached to the master subspace $E$. Through this mapping, the full state vector $\boldsymbol{z}$ is represented as

$$\boldsymbol{z} = \boldsymbol{W}_\varepsilon(\boldsymbol{p}, \phi) = \boldsymbol{W}(\boldsymbol{p}) + \varepsilon \boldsymbol{Y}(\boldsymbol{p}, \phi), \tag{2.16}$$

where $\boldsymbol{W}(\boldsymbol{p})$ and $\boldsymbol{Y}(\boldsymbol{p}, \phi)$ are the autonomous and non-autonomous parametrizations, respectively.

The parametrization (2.16) of the first-order system can be divided as

$$\boldsymbol{z} = \begin{bmatrix} \boldsymbol{w}_\varepsilon(\boldsymbol{p}, \phi) \\ \dot{\boldsymbol{w}}_\varepsilon(\boldsymbol{p}, \phi) \end{bmatrix} = \begin{bmatrix} \boldsymbol{w}(\boldsymbol{p}) + \varepsilon \boldsymbol{y}(\boldsymbol{p}, \phi) \\ \dot{\boldsymbol{w}}(\boldsymbol{p}) + \varepsilon \dot{\boldsymbol{y}}(\boldsymbol{p}, \phi) \end{bmatrix}, \tag{2.17}$$

[1] In case there are multiple phase terms, the circle $\mathbb{S}$ becomes a multi-dimensional torus $\mathbb{T}$.

where $\boldsymbol{w}_\varepsilon(\boldsymbol{p}, \phi)$ describes the displacement field $\boldsymbol{x}$, and $\dot{\boldsymbol{w}}_\varepsilon(\boldsymbol{p}, \phi)$ corresponds to the velocity field $\dot{\boldsymbol{x}}$.

The parametrizations (2.15) and (2.16) are substituted into the full-order system (2.5), leading to the *invariance equation* [4]:

$$\mathbf{B}\left[\frac{\partial \boldsymbol{W}_\varepsilon}{\partial \boldsymbol{p}}\boldsymbol{R}_\varepsilon + \frac{\partial \boldsymbol{W}_\varepsilon}{\partial \phi}\Omega\right] = \mathbf{A}\boldsymbol{W}_\varepsilon + \boldsymbol{F}(\boldsymbol{W}_\varepsilon) + \varepsilon \boldsymbol{G}(\Omega t). \tag{2.18}$$

## 2.3.1 Autonomous SSM Reduction

Under autonomous settings ($\varepsilon = 0$), Eq. (2.18) reduces to the *autonomous invariance equation* [4]:

$$\mathbf{B}\frac{\partial \boldsymbol{W}}{\partial \boldsymbol{p}}\boldsymbol{R} = \mathbf{A}\boldsymbol{W} + \boldsymbol{F}(\boldsymbol{W}). \tag{2.19}$$

**SSM Accuracy**

The accuracy of the SSM reduction [4] can be quantified by looking at the residual of the autonomous invariance equation (2.19):

$$\epsilon_{\text{inv}} = \left\| \mathbf{B}\frac{\partial \boldsymbol{W}}{\partial \boldsymbol{p}}\boldsymbol{R} - \mathbf{A}\boldsymbol{W} - \boldsymbol{F}(\boldsymbol{W}) \right\|,$$

where the value of $\epsilon_{\text{inv}}$ depends on the expansion order of the SSM. Therefore, by looking at $\epsilon_{\text{inv}}$, it is possible to automatically adjust the expansion order to achieve the desired accuracy level in the SSM reduction [13]. This property is included in the optimization framework that will be introduced in Chap. 5.

To solve Eq. (2.19), the autonomous parts of the parametrizations (2.15) and (2.17) are approximated through Taylor expansions up to a prescribed order:

$$\boldsymbol{R}(\boldsymbol{p}) \approx \sum_{\boldsymbol{m}\in\mathbb{N}^2} \boldsymbol{R}_{\boldsymbol{m}}\boldsymbol{p}^{\boldsymbol{m}}, \tag{2.20a}$$

$$\boldsymbol{w}(\boldsymbol{p}) \approx \sum_{\boldsymbol{m}\in\mathbb{N}^2} \boldsymbol{w}_{\boldsymbol{m}}\boldsymbol{p}^{\boldsymbol{m}}, \tag{2.20b}$$

$$\dot{\boldsymbol{w}}(\boldsymbol{p}) \approx \sum_{\boldsymbol{m}\in\mathbb{N}^2} \dot{\boldsymbol{w}}_{\boldsymbol{m}}\boldsymbol{p}^{\boldsymbol{m}}, \tag{2.20c}$$

where the multi-index notation [15] is employed to represent the multivariate polynomial expansions. In particular, for a two-dimensional SSM, each expansion order $m$ is associated with $m+1$ multi-indices.

**Multi-Index Notation**

A multi-index $\boldsymbol{m} \in \mathbb{N}^M$ of order $m = |\boldsymbol{m}|_1$ is an $M$-dimensional vector, where arithmetic operations such as addition and subtraction are performed element-wise. The multi-index notation can be used to define multivariate monomials of order $m$ as

$$\boldsymbol{p}^{\boldsymbol{m}} = p_1^{m_1} \cdots p_M^{m_M}.$$

In this work, we focus on two-dimensional multi-indices, as we are dealing with a two-dimensional SSM [15]. The subscript $(\bullet)_{xy}$ is used to denote quantities associated with the multi-index $\boldsymbol{m} = \{x, y\}$. For example, $\boldsymbol{w}_{32}$ is the manifold coefficient corresponding to $\boldsymbol{m} = \{3, 2\}$.

Moreover, for a multi-index $\boldsymbol{m} = \{x, y\}$, we define the symmetric multi-index $\boldsymbol{m}^s = \{y, x\}$.

In these expressions, the vectors $\boldsymbol{R}_{\boldsymbol{m}} = [R_{\boldsymbol{m}}^1, R_{\boldsymbol{m}}^2]^T \in \mathbb{C}^2$ contain the coefficients of the reduced dynamics. In contrast, $\boldsymbol{w}_{\boldsymbol{m}} \in \mathbb{C}^n$ and $\dot{\boldsymbol{w}}_{\boldsymbol{m}} \in \mathbb{C}^n$ represent the coefficients for the displacement and velocity fields, respectively.

The expanded parametrizations (2.20) are substituted into the autonomous invariance equation (2.19) to obtain a set of *cohomological equations* [15], one for each multi-index $\boldsymbol{m}$ at each order:

$$\mathbf{L}_{\boldsymbol{m}} \boldsymbol{w}_{\boldsymbol{m}} = \boldsymbol{h}_{\boldsymbol{m}}, \tag{2.21}$$

where $\boldsymbol{w}_{\boldsymbol{m}}$ is the unknown manifold coefficient, $\mathbf{L}_{\boldsymbol{m}}$ is the cohomological matrix, and $\boldsymbol{h}_{\boldsymbol{m}}$ is the right-hand side vector that collects the contributions coming from orders lower than $\boldsymbol{m}$. As demonstrated by Thurnher et al. [15], the cohomological equations (2.21) do not depend on coefficients of higher orders, thus they can be solved sequentially, starting from the lowest order. Moreover, the cohomological equations at each order can be solved independently of one another, therefore allowing for parallel computation.

**Autonomous SSM**

We summarize below the main computational steps of the autonomous SSM coefficients [14].

Eigenvalue coefficient:

$$\Lambda_{m} = \boldsymbol{m} \cdot \boldsymbol{\Lambda} = m_1\lambda + m_2\bar{\lambda}.$$

Nonlinear force contribution (in tensor notation):

$$f_{m}^{i} = \sum_{\substack{\boldsymbol{u},\boldsymbol{k}\in\mathbb{N}^2 \\ \boldsymbol{u}+\boldsymbol{k}=\boldsymbol{m}}} T_2^{ijk} w_{\boldsymbol{u}}^{j} w_{\boldsymbol{k}}^{k} + \sum_{\substack{\boldsymbol{u},\boldsymbol{k},\boldsymbol{l}\in\mathbb{N}^2 \\ \boldsymbol{u}+\boldsymbol{k}+\boldsymbol{l}=\boldsymbol{m}}} T_3^{ijkl} w_{\boldsymbol{u}}^{j} w_{\boldsymbol{k}}^{k} w_{\boldsymbol{l}}^{l}.$$

Lower-order manifold contributions:

$$\boldsymbol{V}_{m} = \sum_{j=1}^{2} \sum_{\substack{\boldsymbol{u},\boldsymbol{k}\in\mathbb{N}^2 \\ 1<u,k<m \\ \boldsymbol{u}+\boldsymbol{k}-\boldsymbol{e}_j=\boldsymbol{m}}} \boldsymbol{w}_{\boldsymbol{u}} u_j R_{\boldsymbol{k}}^{j}$$

$$\dot{\boldsymbol{V}}_{m} = \sum_{j=1}^{2} \sum_{\substack{\boldsymbol{u},\boldsymbol{k}\in\mathbb{N}^2 \\ 1<u,k<m \\ \boldsymbol{u}+\boldsymbol{k}-\boldsymbol{e}_j=\boldsymbol{m}}} \dot{\boldsymbol{w}}_{\boldsymbol{u}} u_j R_{\boldsymbol{k}}^{j}.$$

Overall lower-order contribution:

$$\boldsymbol{C}_{m} = -\mathbf{M}\dot{\boldsymbol{V}}_{m} - (\Lambda_{m}\mathbf{M} + \mathbf{C})\,\boldsymbol{V}_{m} - \boldsymbol{f}_{m}.$$

Reduced dynamics coefficient:

$$R_{m}^{j} = \frac{\boldsymbol{\varphi}^T \boldsymbol{C}_{m}}{\Lambda_{m} + \lambda_j + \delta}, \quad j = 1, 2,$$

if $\boldsymbol{m}$ and $j$ satisfy the near-resonance condition [15], 0 otherwise.

Eigenvector coefficient:

$$\boldsymbol{D}_{m}^{j} = -\left[\left(\Lambda_{m} + \lambda_j\right)\mathbf{M} + \mathbf{C}\right]\boldsymbol{\varphi}, \quad j = 1, 2.$$

(continued)

Right-hand side of the cohomological equation:

$$\boldsymbol{h}_{m} = \boldsymbol{C}_{m} + \sum_{j=1}^{2} \boldsymbol{D}_{m}^{j} R_{m}^{j}.$$

Cohomological matrix:

$$\mathbf{L}_{m} = \mathbf{K} + \Lambda_{m}\mathbf{C} + \Lambda_{m}^{2}\mathbf{M}.$$

Manifold velocity coefficient:

$$\dot{\boldsymbol{w}}_{m} = \Lambda_{m}\boldsymbol{w}_{m} + \sum_{j=1}^{2} R_{m}^{j}\boldsymbol{\varphi} + \boldsymbol{V}_{m}.$$

In particular, at the leading order ($m = 1$), there are two multi-indices ($\{1, 0\}$ and $\{0, 1\}$). In these two cases, the coefficients for the reduced dynamics are typically chosen as the complex conjugate eigenvalues of the system:

$$\boldsymbol{R}_{10} = [\lambda, 0]^{T}, \quad \boldsymbol{R}_{01} = [0, \bar{\lambda}]^{T}, \tag{2.22}$$

which leads to the following selection of the manifold coefficients:

$$\boldsymbol{w}_{10} = \boldsymbol{\varphi}, \quad \dot{\boldsymbol{w}}_{10} = \lambda\boldsymbol{\varphi}, \quad \boldsymbol{w}_{01} = \boldsymbol{\varphi}, \quad \dot{\boldsymbol{w}}_{01} = \bar{\lambda}\boldsymbol{\varphi}. \tag{2.23}$$

**Symmetric Multi-Index Property**
For a two-dimensional SSM of the second-order mechanical system (2.1), a symmetry arises [15]: the manifold coefficients associated with symmetric multi-indices are complex conjugate (e.g., $\boldsymbol{w}_{23} = \bar{\boldsymbol{w}}_{32}$). This symmetry holds across all coefficients involved in the SSM computation and can be leveraged to significantly reduce computational effort. Specifically, we only need the coefficients corresponding to multi-indices $\boldsymbol{m} = \{x, y\}$ with $x \geq y$. Then, the corresponding coefficients for $\boldsymbol{m} = \{x, y\}$ with $y < x$ are the complex conjugates of their symmetric counterparts.

The right-hand side $\boldsymbol{h}_{m}$ of the cohomological equation (2.21) depends on the reduced dynamics coefficient $\boldsymbol{R}_{m}$ associated with the multi-index $\boldsymbol{m}$. When near-resonances of the form $\Lambda_{m} \approx \lambda_{j}$ for $j = 1, 2$ occur, the cohomological matrix $\mathbf{L}_{m}$ becomes nearly singular [1], and a non-zero choice for $\boldsymbol{R}_{m}$ is required to solve

Eq. (2.21). There are different possible choices for $\boldsymbol{R_m}$, each one resulting in a different parametrization style [4]. In this work, we follow the *normal-form* style parametrization, which was used by Breunung and Haller [2] to obtain analytic expressions for frequency response and backbone curves.

**Near-Resonance Condition**

For a two-dimensional multi-index $\boldsymbol{m}$, the near-resonance condition [15] is satisfied if $\Lambda_{\boldsymbol{m}} \approx \lambda_j$ for $j = 1, 2$.

Let us write the eigenvalue coefficient $\Lambda_{\boldsymbol{m}}$ as

$$\begin{aligned}\Lambda_{\boldsymbol{m}} &= m_1\lambda + m_2\bar{\lambda}\\ &= (m_1 + m_2)\mathrm{Re}\lambda + \mathrm{i}(m_1 - m_2)\mathrm{Im}\lambda,\end{aligned}$$

where $\mathrm{Re}\lambda_1 = \mathrm{Re}\lambda_2 = \mathrm{Re}\lambda$, and $\mathrm{Im}\lambda_1 = -\mathrm{Im}\lambda_2 = \mathrm{Im}\lambda$.

The near-resonance condition is then written as

$$\begin{aligned}(m_1 + m_2)\mathrm{Re}\lambda &\approx \mathrm{Re}\lambda\\ (m_1 - m_2)\mathrm{Im}\lambda &\approx \mathrm{Im}\lambda_j, \quad j = 1, 2.\end{aligned}$$

In case of $\mathrm{Re}\lambda \neq 0$, the above condition is satisfied only for $\boldsymbol{m} = \{1, 0\}$ and $\boldsymbol{m} = \{0, 1\}$. However, under the assumption of lightly-damped systems, it holds that $|\mathrm{Re}\lambda| \ll |\mathrm{Im}\lambda|$. In this case, we can ignore the first constraint and keep only the second one. In particular, if $j = 1$, the condition becomes $m_1 - m_2 = 1$, while we have $m_1 - m_2 = -1$ if $j = 2$.

This condition is only possible for multi-indices at odd orders such that $|m_1 - m_2| = 1$, meaning that at even orders there are no near-resonances and the coefficients $\boldsymbol{R_m}$ are all zeros. Moreover, at odd orders, there are only two multi-indices such that:

$$\begin{aligned}m_1 - m_2 = +1 &\implies R^1_{\boldsymbol{m}} \neq 0\\ m_1 - m_2 = -1 &\implies R^2_{\boldsymbol{m}} \neq 0.\end{aligned}$$

We denote these two non-zero coefficients as:

$$\begin{aligned}\gamma^1_m &= R^1_{\boldsymbol{m}} \quad \text{if} \quad m_1 - m_2 = +1\\ \gamma^2_m &= R^2_{\boldsymbol{m}} \quad \text{if} \quad m_1 - m_2 = -1,\end{aligned}$$

where, according to the symmetric multi-index property [14], the two coefficients are complex conjugate:

$$\gamma^1_m = \bar{\gamma}^2_m = \gamma_m.$$

In the case of conservative systems, i.e., without damping, the autonomous part of the SSM reduces to the Lyapunov Subcenter Manifold (LS$\mathcal{M}$), from which the conservative backbone curve can be computed [4]. In this case, the real part of the master eigenvalues vanishes, $\text{Re}\lambda = 0$, and the near-resonance condition becomes the inner-resonance condition. The LS$\mathcal{M}$ coefficients can be computed following the same steps described here for the SSM by removing the damping, i.e., by setting the Rayleigh damping coefficients $\alpha_R$ and $\beta_R$ to zero.

### 2.3.2 Non-autonomous SSM Contribution

In non-autonomous settings ($\varepsilon > 0$), parametrizations (2.15) and (2.16) are substituted into the invariance equation (2.18). Then, the *non-autonomous invariance equation* is obtained by collecting the $O(\varepsilon)$ terms [4]:

$$\mathbf{B}\left[\frac{\partial \boldsymbol{W}}{\partial \boldsymbol{p}}\boldsymbol{S} + \frac{\partial \boldsymbol{Y}}{\partial \boldsymbol{p}}\boldsymbol{R} + \frac{\partial \boldsymbol{Y}}{\partial \phi}\Omega\right] = \mathbf{A}\boldsymbol{Y} + \mathcal{D}\boldsymbol{F}(\boldsymbol{W})\boldsymbol{Y} + \boldsymbol{G}, \tag{2.24}$$

where the operator $\mathcal{D}(\bullet)$ represents the total derivative of $(\bullet)$. Equation (2.24) needs to be solved for the non-autonomous terms $\boldsymbol{S}$ and $\boldsymbol{Y}$ in the parametrizations (2.15) and (2.16), respectively.

Using the multi-index notation, the terms $\boldsymbol{S}$ and $\boldsymbol{Y}$ are expanded using a combination of Taylor and Fourier series [15]:

$$\begin{aligned}
\boldsymbol{S}(\boldsymbol{p}, \phi) &\approx \sum_{\kappa\in\mathbb{Z}} \boldsymbol{S}_{0,\kappa} e^{\kappa i\phi} + \sum_{\kappa\in\mathbb{Z}} \sum_{\boldsymbol{m}\in\mathbb{N}^2} \boldsymbol{S}_{\boldsymbol{m},\kappa} e^{\kappa i\phi} \boldsymbol{p}^{\boldsymbol{m}}, \\
\boldsymbol{y}(\boldsymbol{p}, \phi) &\approx \sum_{\kappa\in\mathbb{Z}} \boldsymbol{y}_{0,\kappa} e^{\kappa i\phi} + \sum_{\kappa\in\mathbb{Z}} \sum_{\boldsymbol{m}\in\mathbb{N}^2} \boldsymbol{y}_{\boldsymbol{m},\kappa} e^{\kappa i\phi} \boldsymbol{p}^{\boldsymbol{m}}, \\
\dot{\boldsymbol{y}}(\boldsymbol{p}, \phi) &\approx \sum_{\kappa\in\mathbb{Z}} \dot{\boldsymbol{y}}_{0,\kappa} e^{\kappa i\phi} + \sum_{\kappa\in\mathbb{Z}} \sum_{\boldsymbol{m}\in\mathbb{N}^2} \dot{\boldsymbol{y}}_{\boldsymbol{m},\kappa} e^{\kappa i\phi} \boldsymbol{p}^{\boldsymbol{m}},
\end{aligned} \tag{2.25}$$

where the summation index $\kappa$ represents contributions to distinct harmonics in the Fourier expansion.

Similarly to the autonomous case, the parametrizations (2.25) are substituted into Eq. (2.24), leading to the following cohomological equations, one for every combination of $\boldsymbol{m}$ and $\kappa$ [15]:

$$\mathbf{L}_{\boldsymbol{m},\kappa} \boldsymbol{y}_{\boldsymbol{m},\kappa} = \boldsymbol{h}_{\boldsymbol{m},\kappa}. \tag{2.26}$$

As done by Jain and Haller [4], we focus on the leading-order terms $\boldsymbol{S}_{0,\kappa}$, $\boldsymbol{y}_{0,\kappa}$, and $\dot{\boldsymbol{y}}_{0,\kappa}$, which do not depend on the reduced coordinates $\boldsymbol{p}$. Moreover, the Fourier expansion of the single harmonic external force in (2.1) has only two non-zero terms associated with $\kappa = -1, 1$:

$$\boldsymbol{g}\cos(\phi) = \sum_{\kappa=-1,1} \boldsymbol{g}_{0,\kappa} e^{\kappa i \phi}, \tag{2.27}$$

where the known Fourier coefficients are $\boldsymbol{g}_{0,-1} = \boldsymbol{g}_{0,1} = \boldsymbol{g}/2$. In this settings, Eq. (2.26) simplifies to

$$\mathbf{L}_{0,\kappa} \boldsymbol{y}_{0,\kappa} = \boldsymbol{h}_{0,\kappa}, \tag{2.28}$$

from which the coefficients $\boldsymbol{y}_{0,\kappa}$ are computed.

**Non-autonomous, Leading-Order SSM**

We summarize below the main computational steps of the non-autonomous, leading-order SSM coefficients [14].

Eigenvalue coefficient:

$$\Lambda_{0,\kappa} = \kappa i \Omega.$$

Reduced dynamics coefficient:

$$S^j_{0,\kappa} = \frac{-\boldsymbol{\varphi}^T \boldsymbol{g}_{0,\kappa}}{\Lambda_{0,\kappa} + \lambda_j + \delta}, \quad j = 1, 2,$$

if $\kappa$ and $j$ satisfy the near-resonance condition, 0 otherwise.

Eigenvector coefficient:

$$\boldsymbol{D}^j_{0,\kappa} = -\left[\left(\Lambda_{0,\kappa} + \lambda_j\right)\mathbf{M} + \mathbf{C}\right]\boldsymbol{\varphi}, \quad j = 1, 2.$$

Right-hand side of the cohomological equation:

$$\boldsymbol{h}_{0,\kappa} = -\boldsymbol{g}_{0,\kappa} + \sum_{j=1}^{2} \boldsymbol{D}^j_{0,\kappa} S^j_{0,\kappa}.$$

Cohomological matrix:

$$\mathbf{L}_{0,\kappa} = \mathbf{K} + \Lambda_{0,\kappa}\mathbf{C} + \Lambda^2_{0,\kappa}\mathbf{M}.$$

(continued)

Manifold velocity coefficient:

$$\dot{\boldsymbol{y}}_{0,\kappa} = \boldsymbol{\Lambda}_{0,\kappa}\boldsymbol{y}_{0,\kappa} + \sum_{j=1}^{2} \boldsymbol{S}_{0,\kappa}^{j}\boldsymbol{\varphi}.$$

In case of near-resonances of the form $\kappa \mathrm{i}\Omega \approx \lambda_j$, the coefficient matrix $\mathbf{L}_{0,\kappa}$ becomes nearly singular [4]. In this case, a non-zero choice for $\boldsymbol{S}_{0,\kappa}$ is required to solve Eq. (2.28). Similarly to the autonomous case, we select $\boldsymbol{S}_{0,\kappa}$ following the normal-form style parametrization [4].

**Non-autonomous, Near-Resonance Condition**
Assume a periodic forcing with excitation frequency in near-resonance with the master eigenvalue, $\Omega = |\mathrm{Im}\lambda|$. Then, the near-resonance condition $\kappa \mathrm{i}\Omega \approx \lambda_j$ is satisfied in the following cases [4]:

$$j = 1, \quad \kappa = +1 \implies S_{0,+1}^{1} \neq 0$$
$$j = 2, \quad \kappa = -1 \implies S_{0,-1}^{2} \neq 0.$$

According to the symmetric multi-index property [14], the two coefficients are complex conjugates:

$$S_{0,+1}^{1} = \bar{S}_{0,-1}^{2} = S_0.$$

## 2.4 Periodic Orbits of the Reduced Dynamics

To determine the steady-state response of the system (2.1), we need to identify periodic orbits of the reduced dynamics (2.15). By following the approach described by Thurnher et al. [15], we transform the parametrization vector $\boldsymbol{p}$ into polar coordinates $(\rho, \theta)$:

$$\boldsymbol{p} = \begin{bmatrix} p \\ \bar{p} \end{bmatrix} = \rho \begin{bmatrix} e^{\mathrm{i}\theta} \\ e^{-\mathrm{i}\theta} \end{bmatrix} = \rho \tilde{\boldsymbol{p}}. \tag{2.29}$$

The expression for the reduced dynamics of $p = \rho e^{\mathrm{i}\theta}$ is given by

$$\dot{p} = \sum_{\boldsymbol{m}\in\mathbb{N}^2} R^1_{\boldsymbol{m}} p^{m_1} \bar{p}^{m_2} + \varepsilon \sum_{\kappa=-1,1} S^1_{0,\kappa} e^{\kappa \mathrm{i}\phi}, \tag{2.30}$$

which in polar coordinates reads

$$\left(\dot{\rho} + \mathrm{i}\rho\dot{\theta}\right) e^{\mathrm{i}\theta} = \sum_{\boldsymbol{m}\in\mathbb{N}^2} R^1_{\boldsymbol{m}} \rho^{m} e^{(m_1-m_2)\mathrm{i}\theta} + \varepsilon \sum_{\kappa=-1,1} S^1_{0,\kappa} e^{\kappa \mathrm{i}\phi}. \tag{2.31}$$

Having used the normal-form style parametrization, the only non-zero $R^1_{\boldsymbol{m}}$ coefficients correspond to the multi-indices with $m_1 - m_2 = 1$. The same applies to the coefficient $S^1_{0,\kappa}$, which is non-zero if $\kappa = 1$. Therefore, the above expression is simplified as

$$\left(\dot{\rho} + \mathrm{i}\rho\dot{\theta}\right) e^{\mathrm{i}\theta} = \sum_{\boldsymbol{m}\in\mathbb{N}^2} R^1_{\boldsymbol{m}} \rho^{m} e^{\mathrm{i}\theta} + \varepsilon S^1_{0,1} e^{\mathrm{i}\phi}, \tag{2.32}$$

which is divided by the term $e^{\mathrm{i}\theta}$, leading to

$$\dot{\rho} + \mathrm{i}\rho(\dot{\psi} + \Omega) = \sum_{\boldsymbol{m}\in\mathbb{N}^2} R^1_{\boldsymbol{m}} \rho^{m} + \varepsilon S^1_{0,1} e^{-\mathrm{i}\psi}, \tag{2.33}$$

where $\psi = \theta - \phi$ is the phase shift between the response and the forcing, for which $\dot{\theta} = \dot{\psi} + \Omega$ and $\Omega = \dot{\phi}$.

Finally, by separating the real and imaginary parts, the expression for the reduced dynamics in polar coordinates is given by

$$\begin{bmatrix} \dot{\rho} \\ \rho\dot{\psi} \end{bmatrix} = \boldsymbol{r}(\rho, \psi, \Omega), \tag{2.34a}$$

$$\dot{\phi} = \Omega, \tag{2.34b}$$

where

$$\boldsymbol{r}(\rho, \psi, \Omega) = \begin{bmatrix} a(\rho) \\ b(\rho, \Omega) \end{bmatrix} + \varepsilon \mathbf{N}(\psi) \begin{bmatrix} \mathrm{Re} S^1_{0,1} \\ \mathrm{Im} S^1_{0,1} \end{bmatrix} \tag{2.35a}$$

$$a(\rho) = \sum_{\boldsymbol{m}\in\mathbb{N}^2} \mathrm{Re} R^1_{\boldsymbol{m}} \rho^{m} \tag{2.35b}$$

$$b(\rho, \Omega) = \sum_{\boldsymbol{m}\in\mathbb{N}^2} \mathrm{Im} R^1_{\boldsymbol{m}} \rho^{m} - \rho\Omega = \tilde{b}(\rho) - \rho\Omega \tag{2.35c}$$

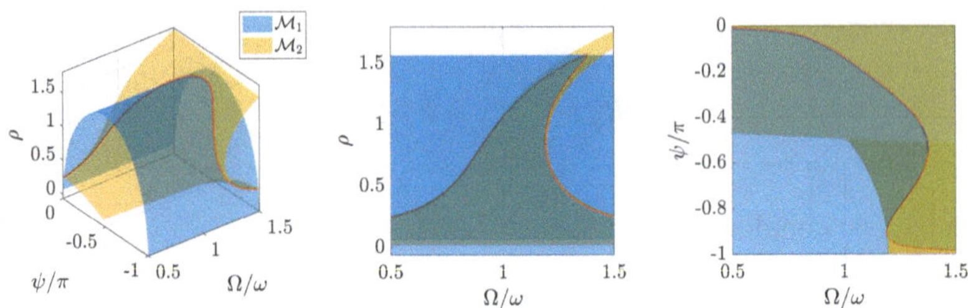


**Fig. 2.1** Geometric representation of the frequency response. The intersection between the zero-level sets $\mathcal{M}_1$ and $\mathcal{M}_2$ of Eq. (2.36) leads to the frequency response (solid red curve)

$$\mathbf{N}(\psi) = \begin{bmatrix} \cos(\psi) & \sin(\psi) \\ -\sin(\psi) & \cos(\psi) \end{bmatrix}. \tag{2.35d}$$

The fixed points of the SSM-reduced dynamics correspond to periodic orbits of the full system, whose amplitude, phase, and frequency satisfy

$$\boldsymbol{r}(\rho, \psi, \Omega) = \begin{bmatrix} r_1(\rho, \psi, \Omega) \\ r_2(\rho, \psi, \Omega) \end{bmatrix} = \mathbf{0}. \tag{2.36}$$

As done by Ponsioen et al. [11], we define the zero-level sets of $r_1$ and $r_2$ as $\mathcal{M}_1$ and $\mathcal{M}_2$, respectively. The frequency-response curve are then given as the intersection of $\mathcal{M}_1$ and $\mathcal{M}_2$ in the $(\rho, \psi, \Omega)$-space, as illustrated in Fig. 2.1.

The amplitude of the periodic response is given by the zeros of the equation

$$\mathcal{F}(\rho, \Omega) = (a(\rho))^2 + (b(\rho, \Omega))^2 - \left(\varepsilon |S_{0,1}^1|\right)^2 = 0, \tag{2.37}$$

or, alternatively, by

$$\mathcal{G}(\rho, \Omega) = b(\rho, \Omega) \pm \sqrt{\left(\varepsilon |S_{0,1}^1|\right)^2 - (a(\rho))^2} = 0, \tag{2.38}$$

while the phase shift $\psi$ is given by

$$\psi = \arctan\left( \frac{b(\rho, \Omega)\mathrm{Re}S_{0,1}^1 - a(\rho)\mathrm{Im}S_{0,1}^1}{-a(\rho)\mathrm{Re}S_{0,1}^1 - b(\rho, \Omega)\mathrm{Im}S_{0,1}^1} \right). \tag{2.39}$$

The eigenvalues of the Jacobian of the SSM-reduced dynamics determine the stability of the fixed points [2]:

$$J(\rho, \Omega) = \begin{bmatrix} \frac{\partial a(\rho)}{\partial \rho} & -b(\rho, \Omega) \\ \frac{1}{\rho} \frac{\partial b(\rho, \Omega)}{\partial \rho} & \frac{1}{\rho} a(\rho) \end{bmatrix}. \tag{2.40}$$

In particular, using the Routh–Hurwitz criterion [2] on the Jacobian (2.40), a fixed point is stable if $\mathrm{tr}(J) < 0$ and $\det(J) > 0$. At bifurcation points, the stability of the response changes, and the two inequality conditions become equality conditions. These bifurcations appear along straight lines in the $\Omega$, $\rho$ space.

## 2.5 Frequency Response and Backbone Curves

Following the definition of Breunung and Haller [2], the *backbone curve* of the mechanical system (2.1) is the *curve of maximal amplitude of the periodic response on the SSM as a function of the frequency of the external forcing.*

In particular, points on the backbone curve can be computed by setting to zero the derivative of Eq. (2.37):

$$\frac{\partial \mathcal{F}}{\partial \Omega} = 2b(\rho, \Omega)\rho = 2\rho(\tilde{b}(\rho) - \rho\Omega). \tag{2.41}$$

Neglecting the trivial solution $\rho = 0$, Eq. (2.37) leads to the following analytical expression that relates the frequency $\Omega_{\max}$ at which the amplitude of the forced response is at a maximum:

$$\Omega_{\max}(\rho_{\max}) = \frac{1}{\rho_{\max}}\tilde{b} = \mathrm{Im}\lambda + \sum_{\substack{\boldsymbol{m}\in\mathbb{N}^2 \\ |\boldsymbol{m}|>1}} \mathrm{Im}\{R^1_{\boldsymbol{m}}\}\rho_{\max}^{m-1}. \tag{2.42}$$

Additionally, along the backbone curve, the phase shift computed from Eq. (2.39) is always $\psi(\rho_{\max}) = \pi/2$ [2].

Equation (2.42) can be used to parametrize the maximum frequency $\Omega_{\max}$ with respect to $\rho_{\max}$, but this is independent from the forcing level $\varepsilon$. To add the dependency on $\varepsilon$, we combine Eqs. (2.41) and (2.37):

$$(a(\rho_{\max}))^2 - \left(\varepsilon|S^1_{0,1}|\right)^2 = 0, \tag{2.43}$$

which can be solved for $\rho_{\max}$ given the value of $\varepsilon$.

Finally, Eq. (2.38) is solved to obtain a parametrization for the frequency response curves in the $(\Omega, \rho)$ space:

$$\Omega(\rho) = \frac{1}{\rho}\tilde{b}(\rho) \pm \frac{1}{\rho}\sqrt{\left(\varepsilon|S^1_{0,1}|\right)^2 - (a(\rho))^2}, \tag{2.44}$$

where only real solutions of $\Omega$ are relevant. Moreover, Eq. (2.44) implies that frequency response curves are symmetric with respect to the backbone curve (Fig. 2.3).

## 2.6 Periodic Response in Physical Coordinates

Periodic orbits in the reduced space $(\Omega, \rho)$ are related to periodic orbits in the original physical coordinates through the manifold parametrization (2.17). Typically, in mechanical problems, we are interested in finding the periodic response of one degree of freedom $x^i$, which can be computed as

$$\begin{aligned} x^i(\rho, \phi) &= x^i_{\text{auto}} + x^i_{\text{nonauto}} \\ &= \sum_{\boldsymbol{m}\in\mathbb{N}^2} w^i_{\boldsymbol{m}} \boldsymbol{p}^{\boldsymbol{m}} + \varepsilon \sum_{\kappa=-1,1} y^i_{0,\kappa} e^{\kappa \mathrm{i}\phi} \end{aligned} \tag{2.45}$$

where $(\bullet)^i$ represents the $i$-th component of the vector $(\bullet)$. The reduced coordinate $\boldsymbol{p}$ depends on the reduced amplitude $\rho$ and on the phase shift $\psi$:

$$\boldsymbol{p} = \rho \begin{bmatrix} e^{\mathrm{i}\theta} \\ e^{-\mathrm{i}\theta} \end{bmatrix} = \rho \begin{bmatrix} e^{\mathrm{i}(\phi+\psi)} \\ e^{-\mathrm{i}(\phi+\psi)} \end{bmatrix}, \tag{2.46}$$

where $\phi = \Omega t$ is responsible for the time dependence of $x^i$. We note that the non-autonomous term in Eq. (2.45) is much smaller than the autonomous term [7], thus leading to the approximation $x^i \approx x^i_{\text{auto}}$.

Using the periodic response of $x^i$, there are different options to define the associated *amplitude*: it could be the maximum value, the amplitude of a spectral component, or the Root Mean Square (RMS) value. In this work, we define the amplitude $x$ associated with the periodic response of $x^i$ using the RMS of the response within one period $T = 2\pi/\Omega$ [13]:

$$x(\rho) = \sqrt{\frac{1}{N_t} \sum_{k=1}^{N_t} \left(x^i(\rho, \phi_k)\right)^2}, \tag{2.47}$$

where $\phi_k = \Omega t_k$ and $t_k = k \cdot T/N_t$ for $k = 0, \ldots, N_t - 1$.

We have now defined all the steps to compute the backbone curve and the frequency response amplitude for the system (2.1). Given a forcing level $\varepsilon$, we can solve Eq. (2.43) to obtain the maximum reduced amplitude $\rho_{\max}$. Then, we can use Eqs. (2.42) and (2.47) to compute the response frequency $\Omega_{\max}$ and the physical amplitude $x_{\max}$ at the peak of the frequency response.

All the routines described in this chapter are implemented using `YetAnotherFE code` [6], an open-source MATLAB-based code that implements the Finite Element Method (FEM) in an object-oriented fashion, including the tensor notation used to describe the quadratic and cubic nonlinearities (2.3).

## 2.7 Numerical Example

Consider, as an example, the model of a geometrically nonlinear, clamped-clamped, uniform oscillator chain (Fig. 2.2). The system has 7 masses connected by linear, quadratic, and cubic springs (Table 2.1).

The equation of motion for the $n$-mass oscillator chain is given by Eq. (2.1) with

$$\begin{aligned} \mathbf{M} &= m\mathbf{I}_n, \quad \mathbf{C} = c\mathbf{L}_n, \quad \mathbf{K} = k\mathbf{L}_n, \\ \boldsymbol{f}(\boldsymbol{x}) &= \boldsymbol{f}_2(\boldsymbol{x}, \boldsymbol{x}) + \boldsymbol{f}_3(\boldsymbol{x}, \boldsymbol{x}, \boldsymbol{x}), \end{aligned} \tag{2.48}$$

where $\mathbf{I}_n \in \mathbb{R}^{n \times n}$ is the $n$-dimensional identity matrix and $\mathbf{L}_n \in \mathbb{R}^{n \times n}$ is a Toeplitz matrix given as

$$\mathbf{L}_n = \begin{bmatrix} 2 & -1 & & & \\ -1 & 2 & -1 & & \\ & \ddots & \ddots & \ddots & \\ & & -1 & 2 & -1 \\ & & & -1 & 2 \end{bmatrix}. \tag{2.49}$$

The quadratic and cubic nonlinearities are defined using the sparse tensor notation such that

$$\boldsymbol{f}_2(\boldsymbol{x}, \boldsymbol{x}) = k_2 \begin{bmatrix} x_1^2 - (x_2 - x_1)^2 \\ (x_2 - x_1)^2 - (x_3 - x_2)^2 \\ \vdots \\ (x_n - x_{n-1})^2 - x_n^2 \end{bmatrix} \tag{2.50a}$$

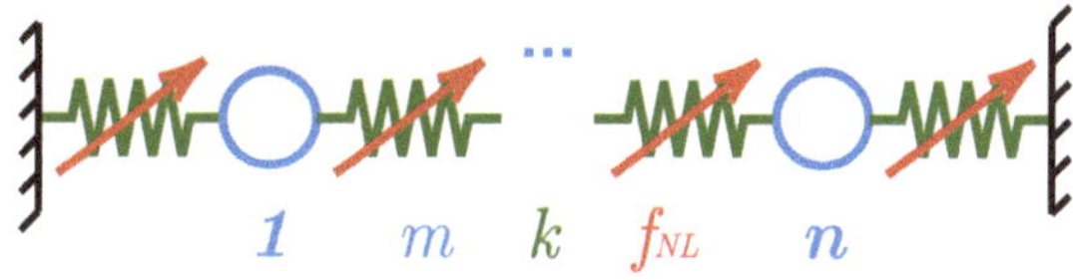


**Fig. 2.2** Model of a nonlinear, clamped-clamped, uniform oscillator chain

**Table 2.1** Parameters of the nonlinear, clamped-clamped, uniform oscillator chain

| | Symbol | Value |
|---|---|---|
| Mass | $m$ | 1.0 kg |
| Linear damping | $c$ | 0.1 Ns/m |
| Linear spring | $k$ | 1.0 N/m |
| Quadratic spring | $k_2$ | 0.5 N/m$^2$ |
| Cubic spring | $k_3$ | 0.3 N/m$^3$ |

$$\boldsymbol{f}_3(\boldsymbol{x},\boldsymbol{x},\boldsymbol{x}) = k_3 \begin{bmatrix} x_1^3 - (x_2 - x_1)^3 \\ (x_2 - x_1)^3 - (x_3 - x_2)^3 \\ \vdots \\ (x_n - x_{n-1})^3 - x_n^3 \end{bmatrix}. \tag{2.50b}$$

The SSM is built on the master subspace spanned by the first mode shape of the system, with an external forcing frequency in near-resonance with the master eigenvalue $\lambda = -0.0076 + \mathrm{i}0.3901$. The external force is applied to the middle mass, and the frequency response amplitude is measured at the same location.

The $O(5)$ autonomous SSM is shown in Fig. 2.3, along with the frequency response and backbone curves evaluated at different forcing levels ($\varepsilon = 0.01, 0.1, 0.15$ N). Additionally, the stability of the response with the highest forcing level is analyzed.

For reference, Tables 2.2, 2.3, and 2.4 summarize the computation steps of the autonomous SSM up to the third order.

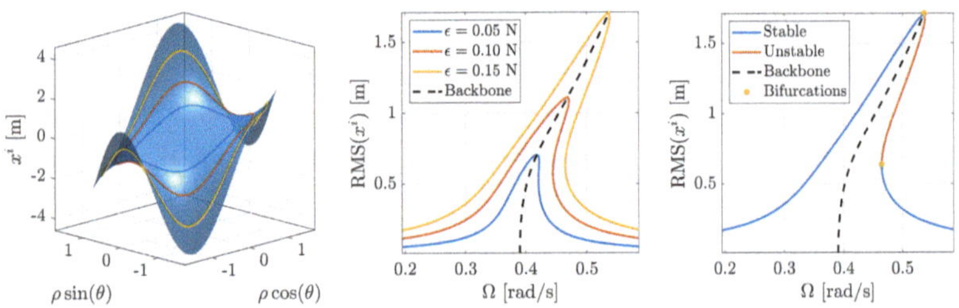


**Fig. 2.3** Periodic $O(5)$ response of the nonlinear, clamped-clamped, uniform oscillator chain. The periodic orbits on the autonomous SSM (left) correspond to frequency response curves at different forcing levels (middle). Additionally, the stability analysis is performed on the highest frequency response (right)

**Table 2.2** SSM computations at leading order ($|\boldsymbol{m}|_1 = 1$)

| $\boldsymbol{m} = \{1, 0\}$ | $\boldsymbol{m} = \{0, 1\}$ |
|---|---|
| $\boldsymbol{R}_{10} = \begin{bmatrix} \lambda \\ 0 \end{bmatrix}$ | $\boldsymbol{R}_{01} = \begin{bmatrix} 0 \\ \bar{\lambda} \end{bmatrix}$ |
| $\boldsymbol{w}_{10} = \boldsymbol{\varphi}$ | $\boldsymbol{w}_{01} = \boldsymbol{\varphi}$ |
| $\dot{\boldsymbol{w}}_{10} = \lambda\boldsymbol{\varphi}$ | $\dot{\boldsymbol{w}}_{01} = \bar{\lambda}\boldsymbol{\varphi}$ |

**Table 2.3** SSM computations at second order ($|\boldsymbol{m}|_1 = 2$)

| $\boldsymbol{m} = \{2, 0\}$ | $\boldsymbol{m} = \{0, 2\}$ |
|---|---|
| $\Lambda_{20} = 2\lambda$<br>$\boldsymbol{f}_{20} = \boldsymbol{f}_2(\boldsymbol{w}_{10}, \boldsymbol{w}_{10})$<br>$\boldsymbol{R}_{20} = \mathbf{0}$<br>$\mathbf{L}_{20} = \mathbf{K} + \Lambda_{20}\mathbf{C} + \Lambda_{20}^2\mathbf{M}$<br>$\boldsymbol{h}_{20} = -\boldsymbol{f}_{20}$<br>$\mathbf{L}_{20}\boldsymbol{w}_{20} - \boldsymbol{h}_{20} = \mathbf{0}$<br>$\dot{\boldsymbol{w}}_{20} = \Lambda_{20}\boldsymbol{w}_{20}$ | $\Lambda_{02} = 2\bar{\lambda}$<br>$\boldsymbol{f}_{02} = \boldsymbol{f}_2(\boldsymbol{w}_{01}, \boldsymbol{w}_{01})$<br>$\boldsymbol{R}_{02} = \mathbf{0}$<br>$\mathbf{L}_{02} = \mathbf{K} + \Lambda_{02}\mathbf{C} + \Lambda_{02}^2\mathbf{M}$<br>$\boldsymbol{h}_{02} = -\boldsymbol{f}_{02}$<br>$\mathbf{L}_{02}\boldsymbol{w}_{02} - \boldsymbol{h}_{02} = \mathbf{0}$<br>$\dot{\boldsymbol{w}}_{02} = \Lambda_{02}\boldsymbol{w}_{02}$ |
| $\boldsymbol{m} = \{1, 1\}$ | |
| $\Lambda_{11} = 2\mathrm{Re}\lambda$<br>$\boldsymbol{f}_{11} = \boldsymbol{f}_2(\boldsymbol{w}_{10}, \boldsymbol{w}_{01})$<br>$\quad + \boldsymbol{f}_2(\boldsymbol{w}_{01}, \boldsymbol{w}_{10})$<br>$\boldsymbol{R}_{11} = \mathbf{0}$<br>$\mathbf{L}_{11} = \mathbf{K} + \Lambda_{11}\mathbf{C} + \Lambda_{11}^2\mathbf{M}$<br>$\boldsymbol{h}_{11} = -\boldsymbol{f}_{11}$<br>$\mathbf{L}_{11}\boldsymbol{w}_{11} - \boldsymbol{h}_{11} = \mathbf{0}$<br>$\dot{\boldsymbol{w}}_{11} = \Lambda_{11}\boldsymbol{w}_{11}$ | |

**Table 2.4** SSM computations at third order ($|\boldsymbol{m}|_1 = 3$)

| $\boldsymbol{m} = \{3, 0\}$ | $\boldsymbol{m} = \{0, 3\}$ |
|---|---|
| $\Lambda_{30} = 3\lambda$<br>$\boldsymbol{f}_{30} = \boldsymbol{f}_2(\boldsymbol{w}_{20}, \boldsymbol{w}_{10})$<br>$\quad + \boldsymbol{f}_2(\boldsymbol{w}_{10}, \boldsymbol{w}_{20})$<br>$\quad + \boldsymbol{f}_3(\boldsymbol{w}_{10}, \boldsymbol{w}_{10}, \boldsymbol{w}_{10})$<br>$\boldsymbol{R}_{30} = \mathbf{0}$<br>$\mathbf{L}_{30} = \mathbf{K} + \Lambda_{30}\mathbf{C} + \Lambda_{30}^2\mathbf{M}$<br>$\boldsymbol{h}_{30} = -\boldsymbol{f}_{30}$<br>$\mathbf{L}_{30}\boldsymbol{w}_{30} - \boldsymbol{h}_{30} = \mathbf{0}$<br>$\dot{\boldsymbol{w}}_{30} = \Lambda_{30}\boldsymbol{w}_{30}$ | $\Lambda_{03} = 3\bar{\lambda}$<br>$\boldsymbol{f}_{03} = \boldsymbol{f}_2(\boldsymbol{w}_{02}, \boldsymbol{w}_{01})$<br>$\quad + \boldsymbol{f}_2(\boldsymbol{w}_{01}, \boldsymbol{w}_{02})$<br>$\quad + \boldsymbol{f}_3(\boldsymbol{w}_{01}, \boldsymbol{w}_{01}, \boldsymbol{w}_{01})$<br>$\boldsymbol{R}_{03} = \mathbf{0}$<br>$\mathbf{L}_{03} = \mathbf{K} + \Lambda_{03}\mathbf{C} + \Lambda_{03}^2\mathbf{M}$<br>$\boldsymbol{h}_{03} = -\boldsymbol{f}_{03}$<br>$\mathbf{L}_{03}\boldsymbol{w}_{03} - \boldsymbol{h}_{03} = \mathbf{0}$<br>$\dot{\boldsymbol{w}}_{03} = \Lambda_{03}\boldsymbol{w}_{03}$ |

(continued)

**Table 2.4** (continued)

| $\boldsymbol{m} = \{2, 1\}$ | $\boldsymbol{m} = \{1, 2\}$ |
|---|---|
| $\Lambda_{21} = 2\lambda + \bar{\lambda}$ | $\Lambda_{12} = \lambda + 2\bar{\lambda}$ |
| $\boldsymbol{f}_{21} = \boldsymbol{f}_2(\boldsymbol{w}_{10}, \boldsymbol{w}_{11}) + \boldsymbol{f}_2(\boldsymbol{w}_{11}, \boldsymbol{w}_{10}) + \boldsymbol{f}_2(\boldsymbol{w}_{01}, \boldsymbol{w}_{20}) + \boldsymbol{f}_2(\boldsymbol{w}_{20}, \boldsymbol{w}_{01}) + \boldsymbol{f}_3(\boldsymbol{w}_{10}, \boldsymbol{w}_{10}, \boldsymbol{w}_{01}) + \boldsymbol{f}_3(\boldsymbol{w}_{10}, \boldsymbol{w}_{01}, \boldsymbol{w}_{10}) + \boldsymbol{f}_3(\boldsymbol{w}_{01}, \boldsymbol{w}_{10}, \boldsymbol{w}_{10})$ | $\boldsymbol{f}_{12} = \boldsymbol{f}_2(\boldsymbol{w}_{01}, \boldsymbol{w}_{11}) + \boldsymbol{f}_2(\boldsymbol{w}_{11}, \boldsymbol{w}_{01}) + \boldsymbol{f}_2(\boldsymbol{w}_{10}, \boldsymbol{w}_{02}) + \boldsymbol{f}_2(\boldsymbol{w}_{02}, \boldsymbol{w}_{10}) + \boldsymbol{f}_3(\boldsymbol{w}_{10}, \boldsymbol{w}_{01}, \boldsymbol{w}_{01}) + \boldsymbol{f}_3(\boldsymbol{w}_{01}, \boldsymbol{w}_{10}, \boldsymbol{w}_{01}) + \boldsymbol{f}_3(\boldsymbol{w}_{01}, \boldsymbol{w}_{01}, \boldsymbol{w}_{10})$ |
| $\boldsymbol{R}_{21} = \begin{bmatrix} \gamma_3 \\ 0 \end{bmatrix}$ | $\boldsymbol{R}_{12} = \begin{bmatrix} 0 \\ \bar{\gamma}_3 \end{bmatrix}$ |
| $\gamma_3 = \dfrac{-\boldsymbol{\varphi}^T \boldsymbol{f}_{21}}{\Lambda_{21} + \lambda + \delta}$ | $\bar{\gamma}_3 = \dfrac{-\boldsymbol{\varphi}^T \boldsymbol{f}_{12}}{\Lambda_{12} + \bar{\lambda} + \delta}$ |
| $\mathbf{L}_{21} = \mathbf{K} + \Lambda_{21}\mathbf{C} + \Lambda_{21}^2\mathbf{M}$ | $\mathbf{L}_{12} = \mathbf{K} + \Lambda_{12}\mathbf{C} + \Lambda_{12}^2\mathbf{M}$ |
| $\boldsymbol{D}_{21}^1 = -\left[(\Lambda_{21} + \lambda)\mathbf{M} + \mathbf{C}\right]\boldsymbol{\varphi}$ | $\boldsymbol{D}_{12}^2 = -\left[(\Lambda_{12} + \bar{\lambda})\mathbf{M} + \mathbf{C}\right]\boldsymbol{\varphi}$ |
| $\boldsymbol{h}_{21} = \boldsymbol{D}_{21}^2\gamma_3 - \boldsymbol{f}_{21}$ | $\boldsymbol{h}_{12} = \boldsymbol{D}_{12}^2\bar{\gamma}_3 - \boldsymbol{f}_{12}$ |
| $\mathbf{L}_{21}\boldsymbol{w}_{21} - \boldsymbol{h}_{21} = \mathbf{0}$ | $\mathbf{L}_{12}\boldsymbol{w}_{12} - \boldsymbol{h}_{12} = \mathbf{0}$ |
| $\dot{\boldsymbol{w}}_{21} = \Lambda_{21}\boldsymbol{w}_{21} + \gamma_3\boldsymbol{\varphi}$ | $\dot{\boldsymbol{w}}_{12} = \Lambda_{12}\boldsymbol{w}_{12} + \bar{\gamma}_3\boldsymbol{\varphi}$ |

## References


1. Àlex H, Canadell M, Figueras JL, Luque A, Mondelo JM (2016) The parameterization method for invariant manifolds, vol. 195. Springer, Berlin. https://doi.org/10.1007/978-3-319-29662-3
2. Breunung T, Haller G (2018) Explicit backbone curves from spectral submanifolds of forced-damped nonlinear mechanical systems. Proc R Soc A Math Phys Eng Sci 474:20180083. https://doi.org/10.1098/rspa.2018.0083
3. Haller G, Ponsioen S (2016) Nonlinear normal modes and spectral submanifolds: existence, uniqueness and use in model reduction. Nonlinear Dyn 86:1493–1534. https://doi.org/10.1007/s11071-016-2974-z
4. Jain S, Haller G (2022) How to compute invariant manifolds and their reduced dynamics in high-dimensional finite element models. Nonlinear Dyn 107:1417–1450. https://doi.org/10.1007/s11071-021-06957-4
5. Jain S, Li M, Thurnher T, Haller G (2024) SSMTool 2.6: Computation of invariant manifolds in high-dimensional mechanics problems (v2.6). Zenodo. https://doi.org/10.5281/zenodo.13909795
6. Jain S, Marconi J, Pozzi M, Tiso P (2026) YetAnotherFEcode (v1.5.0). Zenodo. https://doi.org/10.5281/zenodo.15257966
7. Li M (2024) Explicit sensitivity analysis of spectral submanifolds of mechanical systems. Nonlinear Dyn 112:16733–16770. https://doi.org/10.1007/s11071-024-09947-4

8. Li M, Haller G (2022) Nonlinear analysis of forced mechanical systems with internal resonance using spectral submanifolds, part ii: bifurcation and quasi-periodic response. Nonlinear Dyn 110:1045–1080. https://doi.org/10.1007/s11071-022-07476-6
9. Li M, Jain S, Haller G (2022) Nonlinear analysis of forced mechanical systems with internal resonance using spectral submanifolds, part i: periodic response and forced response curve. Nonlinear Dyn 110:1005–1043. https://doi.org/10.1007/s11071-022-07714-x
10. Marconi J, Tiso P, Braghin F (2020) A nonlinear reduced order model with parametrized shape defects. Comput Methods Appl Mech Eng 360:112785. https://doi.org/10.1016/j.cma.2019.112785
11. Ponsioen S, Pedergnana T, Haller G (2019) Analytic prediction of isolated forced response curves from spectral submanifolds. Nonlinear Dyn 98:2755–2773. https://doi.org/10.1007/s11071-019-05023-4
12. Ponsioen S, Jain S, Haller G (2020) Model reduction to spectral submanifolds and forced-response calculation in high-dimensional mechanical systems. J Sound Vib 488:115640. https://doi.org/10.1016/j.jsv.2020.115640
13. Pozzi M, Marconi J, Jain S, Braghin F (2024) Backbone curve tailoring via lyapunov subcenter manifold optimization. Nonlinear Dyn 112:15719–15739. https://doi.org/10.1007/s11071-024-09881-5
14. Pozzi M, Marconi J, Jain S, Li M, Braghin F (2025) Adjoint sensitivities for the optimization of nonlinear structural dynamics via spectral submanifolds. Proc R Soc A Math Phys Eng Sci 481:20250244. https://doi.org/10.1098/rspa.2025.0244
15. Thurnher T, Haller G, Jain S (2024) Nonautonomous spectral submanifolds for model reduction of nonlinear mechanical systems under parametric resonance. Chaos Interdiscip J Nonlinear Sci 34:73127. https://doi.org/10.1063/5.0168431

# Chapter 3
# Structural Optimization for MEMS Devices

**Abstract** This chapter reviews optimization techniques commonly used in the design of MEMS devices, with a focus on size optimization due to its widespread industrial adoption. A case study on a single-axis MEMS gyroscope illustrates the optimization process, detailing the modeling approach, sensitivity analysis, and design objectives. The nonlinear dynamic response of the optimized gyroscope is then evaluated using the SSM framework.

## 3.1 Introduction

Optimization techniques have been widely used by industries and companies for the design of MEMS devices, including resonators, accelerometers, and gyroscopes [8]. Among these, *size optimization* is the most commonly used method for structural optimization in industrial MEMS design [4]. In this approach, the device is modeled as an assembly of basic components (masses, beams, connections), whose geometries are described by a set of design variables $\boldsymbol{\mu}$. These variables are then optimized using standard optimization strategies, including parametric optimization [6, 7], shape optimization [16], and stochastic or evolutionary algorithms [3, 17, 18].

Focusing on MEMS gyroscopes, the typical optimization problem aims at imposing the operation frequencies of the device corresponding to the drive and sense modes. In particular, the drive frequency $\omega_d$ is typically chosen in the $20-30$ kHz range, while the frequency mismatch $\Delta\omega$ between the drive $\omega_d$ and sense $\omega_s$ frequencies is usually around 2–10% of $\omega_s$ [1]. Moreover, to avoid undesired vibration modes in the device, we want to place the frequencies $\omega_{sp}$ of the spurious modes as far as possible from the operational frequencies (Fig. 3.1).

 
M. Pozzi, F. Braghin, *Structural Optimization for Nonlinear Dynamics*,
PoliMI SpringerBriefs, https://doi.org/10.1007/978-3-032-22600-6_3

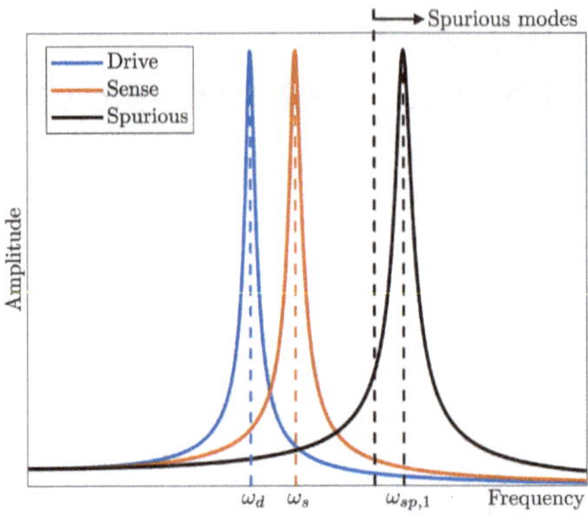


**Fig. 3.1** Typical design requirements for MEMS gyroscopes [7]

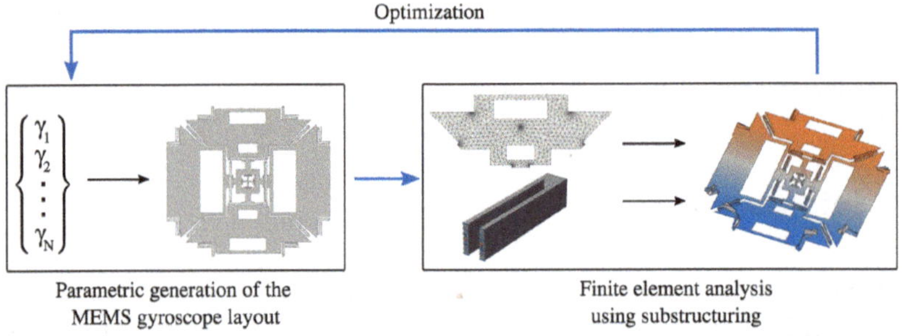


**Fig. 3.2** Working scheme of feMEMS, an automatic design tool for MEMS gyroscopes [7]

Therefore, a typical optimization problem is defined as

$$\begin{aligned} \max_{\boldsymbol{\mu}} \quad & \omega_{sp} \\ \text{s.t.} \quad & \omega_d - \omega_{d,\text{target}} = 0 \\ & \omega_s - \omega_{s,\text{target}} = 0 \\ & \boldsymbol{\mu}_\text{l} \leq \boldsymbol{\mu} \leq \boldsymbol{\mu}_\text{u}, \end{aligned} \tag{3.1}$$

where $\omega_{d,\text{target}}$ and $\omega_{s,\text{target}}$ are the target drive and sense frequencies, and $\boldsymbol{\mu}_\text{l}$ and $\boldsymbol{\mu}_\text{u}$ are the upper and lower bounds for the design variables $\boldsymbol{\mu}$.

In the following sections, we present an optimization case study for a single-axis MEMS gyroscope, detailing the device modeling, sensitivity analysis, and optimization procedure. The study follows the optimization framework of *feMEMS* [6], an automated design tool for MEMS gyroscopes (Fig. 3.2). The same modeling approach and optimization scheme will also be adopted in Chap. 5, where we extend the framework to optimize the nonlinear dynamic response using the Spectral Submanifold (SSM) theory described in Chap. 2.

## 3.2 Gyroscope Model and Analysis

Consider the single-axis MEMS gyroscope designed by Marconi et al. [13]. As done in [6, 7, 14], the layout (Fig. 3.3) is represented as an assembly of rigid bodies, the red drive frame and the blue proof mass, interconnected by elastic suspensions modeled using von Kármán beam elements. Having a two-dimensional problem, each rigid body is characterized by three degrees of freedom, located at its center of mass, along with three lumped inertial properties (mass and moment of inertia with respect to the in-plane rotation).

To enforce compatibility between the rigid bodies and the surrounding elastic structure, we adopt the Multi-Point Constraint (MPC) methodology [5]. In this formulation, each rigid body (*primary*) is coupled to a specific point on the beam mesh (*secondary*) through a linearized kinematic constraint:

$$\begin{bmatrix} u \\ v \\ \theta \end{bmatrix}_{\text{secondary}} = \begin{bmatrix} 1 & 0 & -\Delta y \\ 0 & 1 & \Delta x \\ 0 & 0 & 1 \end{bmatrix} \begin{bmatrix} u \\ v \\ \theta \end{bmatrix}_{\text{primary}}, \tag{3.2}$$

where $u$, $v$, and $\theta$ denote the translational displacements along the $x$- and $y$-axes and the in-plane rotation, respectively. The terms $\Delta x = x_{\text{secondary}} - x_{\text{primary}}$ and $\Delta y = y_{\text{secondary}} - y_{\text{primary}}$ define the relative position of the secondary node with respect to the primary node.

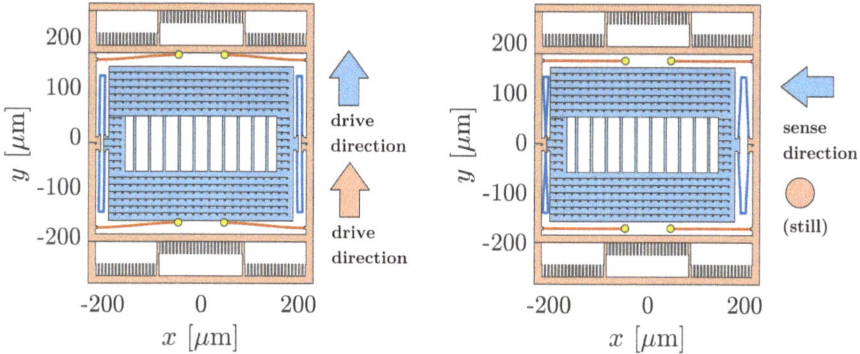


**Fig. 3.3** Drive (left) and sense (right) modes of the initial MEMS gyroscope layout [13]. The drive frame (light red) is connected to the sense mass (light blue) via folded beams (blue), enabling relative motion between the two. The drive frame is anchored to the substrate through drive beams (red), which are attached to fixed supports (yellow circles). Arrows indicate the dominant displacement directions of the drive frame and sense mass in each mode

To illustrate the MPC assembly procedure, let us consider the construction of the mass matrix. For each rigid body, we define a diagonal, lumped mass matrix as

$$\mathbf{M}_j = \begin{bmatrix} m_j & & \\ & m_j & \\ & & J_{zz,j} \end{bmatrix}, \tag{3.3}$$

where $m_j$ denotes the mass of the $j$-th body and $J_{zz,j}$ its moment of inertia about the out-of-plane $z$-axis. By assembling all such contributions into a block-diagonal matrix

$$\mathbf{M}_{\text{lump}} = \begin{bmatrix} \ddots & & \\ & \mathbf{M}_j & \\ & & \ddots \end{bmatrix}, \tag{3.4}$$

and denoting by $\mathbf{M}_b$ the mass matrix associated with the remaining (disconnected) beam elements, the total mass matrix of the system accounting for the MPC constraints is written as

$$\mathbf{M} = \mathbf{T}^T \begin{bmatrix} \mathbf{M}_b & \\ & \mathbf{M}_{\text{lump}} \end{bmatrix} \mathbf{T}, \tag{3.5}$$

where $\mathbf{T}$ is a transformation matrix that enforces the kinematic constraints defined in Eq. (3.2). This matrix effectively eliminates the secondary degrees of freedom and couples the rigid bodies to the beam structures. A similar procedure is used to construct the global stiffness matrix $\mathbf{K}$.

Finally, we can perform the linear modal analysis to evaluate the eigenfrequencies $\omega$ and mode shapes $\boldsymbol{\varphi}$ of the MEMS device:

$$\left(\mathbf{K} - \omega^2 \mathbf{M}\right) \boldsymbol{\varphi} = \mathbf{0}, \tag{3.6}$$

where the mode shape $\boldsymbol{\varphi}$ is assumed to be mass-normalized:

$$\boldsymbol{\varphi}^T \mathbf{M} \boldsymbol{\varphi} = 1. \tag{3.7}$$

### 3.2.1 Sensitivity Analysis

In structural optimization, the sensitivities of the cost function and constraints with respect to the design variables are essential for solving optimization problems using gradient-based methods, such as interior-point and dual-simplex algorithms [4]. These sensitivities quantify how variations in the design parameters affect the values of the objective function and constraints, thus guiding the optimization process toward feasible and optimal solutions. Sensitivities are typically computed using either the *direct differentiation method* or the *adjoint method* [12], depending on the relative number of design variables and constraints.

Consider a scalar function $J(\boldsymbol{\mu}, \boldsymbol{x}) \in \mathbb{R}$, where $\boldsymbol{\mu} \in \mathbb{R}^m$ denotes the vector of design variables and $\boldsymbol{x} \in \mathbb{R}^n$ the vector of state variables, implicitly defined by the state equation $\boldsymbol{f}(\mu, \boldsymbol{x}) = \mathbf{0} \in \mathbb{R}^n$. The direct method is relatively straightforward to implement, as it involves applying the chain rule to differentiate all quantities contributing to $J$. Differentiating the state equation yields:

$$\frac{\partial \boldsymbol{f}}{\partial \mu} + \frac{\partial \boldsymbol{f}}{\partial \boldsymbol{x}} \frac{\mathrm{d}\boldsymbol{x}}{\mathrm{d}\mu} = 0, \quad \forall \mu, \tag{3.8}$$

which can be solved for $\mathrm{d}\boldsymbol{x}/\mathrm{d}\mu$. Then, the total derivative of $J$ with respect to $\mu$ is

$$\frac{\mathrm{d}J}{\mathrm{d}\mu} = \frac{\partial J}{\partial \mu} + \frac{\partial J}{\partial \boldsymbol{x}} \frac{\mathrm{d}\boldsymbol{x}}{\mathrm{d}\mu}, \quad \forall \mu. \tag{3.9}$$

The main drawback of this approach is its computational cost: Eq. (3.8) must be solved for every design variable. As a result, the direct method scales linearly with the number of design parameters, which makes it computationally prohibitive for large-scale problems involving a high-dimensional design space.

To address the scalability limitations of the direct method, the adjoint method offers a more efficient alternative. Instead of computing $\mathrm{d}\boldsymbol{x}/\mathrm{d}\mu$ explicitly for each design parameter, the adjoint method introduces an auxiliary variable, called the *adjoint variable*, to bypass this computation. Starting from $J$ and $\boldsymbol{f}$, we define the *Lagrangian function* as

$$\mathcal{L} = J + \boldsymbol{\lambda}^T \boldsymbol{f}, \tag{3.10}$$

where $\boldsymbol{\lambda} \in \mathbb{R}^n$ is the adjoint variable. Next, we take the partial derivative of $\mathcal{L}$ with respect to $\boldsymbol{x}$, leading to the *adjoint equation*

$$\frac{\partial J}{\partial \boldsymbol{x}} + \boldsymbol{\lambda}^T \frac{\partial \boldsymbol{f}}{\partial \boldsymbol{x}} = \mathbf{0}^T, \tag{3.11}$$

which is solved for $\boldsymbol{\lambda}$. Finally, the sensitivity of $J$ with respect to $\mu$ is equal to the partial derivative of $\mathcal{L}$ with respect to $\mu$:

$$\frac{\mathrm{d}J}{\mathrm{d}\mu} = \frac{\partial \mathcal{L}}{\partial \mu} = \frac{\partial J}{\partial \mu} + \boldsymbol{\lambda}^T \frac{\partial \boldsymbol{f}}{\partial \mu}, \quad \forall \mu. \tag{3.12}$$

While the term $\frac{\partial \mathcal{L}}{\partial \mu}$ must be evaluated for every design variable, the key advantage of this approach is that the adjoint equation is solved only once, independently of the number of design variables, making the adjoint method highly efficient for large problems, including topology optimization [2].

### 3.2.2 Eigenfrequency Sensitivity

We now present how to compute the sensitivities of the eigenfrequency $\omega$ using both the direct approach and the adjoint method. This sensitivity is essential for solving the optimization problem (3.1), as both the objective function and the constraints depend explicitly on the system's eigenfrequencies.

The direct approach involves applying the chain rule to differentiate all quantities contributing to $\omega$. Differentiating the generalized eigenvalue problem (3.6) with respect to a design variable $\mu$ yields:

$$\left(\frac{\mathrm{d}\mathbf{K}}{\mathrm{d}\mu} - \omega^2 \frac{\mathrm{d}\mathbf{M}}{\mathrm{d}\mu} - 2\omega \frac{\mathrm{d}\omega}{\mathrm{d}\mu}\mathbf{M}\right)\boldsymbol{\varphi} + \left(\mathbf{K} - \omega^2\mathbf{M}\right)\frac{\mathrm{d}\boldsymbol{\varphi}}{\mathrm{d}\mu} = \mathbf{0}. \tag{3.13}$$

Additionally, the derivative of the mass-normalization condition (3.7) provides a second equation:

$$2\boldsymbol{\varphi}^T\mathbf{M}\frac{\mathrm{d}\boldsymbol{\varphi}}{\mathrm{d}\mu} + \boldsymbol{\varphi}^T\frac{\mathrm{d}\mathbf{M}}{\mathrm{d}\mu}\boldsymbol{\varphi} = 0. \tag{3.14}$$

By combining Eqs. (3.13) and (3.14), we obtain the following linear system:

$$\begin{bmatrix} \mathbf{K} - \omega^2\mathbf{M} & -2\omega\mathbf{M}\boldsymbol{\varphi} \\ -2\omega\boldsymbol{\varphi}^T\mathbf{M} & 0 \end{bmatrix} \begin{bmatrix} \frac{\mathrm{d}\boldsymbol{\varphi}}{\mathrm{d}\mu} \\ \frac{\mathrm{d}\omega}{\mathrm{d}\mu} \end{bmatrix} = \begin{bmatrix} -\left(\frac{\mathrm{d}\mathbf{K}}{\mathrm{d}\mu} - \omega^2\frac{\mathrm{d}\mathbf{M}}{\mathrm{d}\mu}\right)\boldsymbol{\varphi} \\ \omega\boldsymbol{\varphi}^T\frac{\mathrm{d}\mathbf{M}}{\mathrm{d}\mu}\boldsymbol{\varphi} \end{bmatrix}, \tag{3.15}$$

which can be solved for each design variable $\mu$. Note that this approach requires obtaining the sensitivities of both $\mathrm{d}\omega/\mathrm{d}\mu$ and $\mathrm{d}\boldsymbol{\varphi}/\mathrm{d}\mu$.

While conceptually straightforward, the computational cost of the direct differentiation method scales linearly with the number of design variables, which can become prohibitive in large-scale problems. To overcome this limitation, we adopt the adjoint method, which significantly reduces the computational burden when dealing with many design variables.

We begin by identifying the two state equations, i.e., the generalized eigenvalue problem (3.6) and the mass-normalization condition (3.7). Since we are interested in computing the sensitivity of $\omega$, we select $J = \omega$. Then, using the adjoint variables $\boldsymbol{\lambda}_\varphi$ and $\lambda_\omega$, the Lagrangian is assembled as

$$\mathcal{L} = \omega + \boldsymbol{\lambda}_\varphi^T\left(\mathbf{K} - \omega^2\mathbf{M}\right)\boldsymbol{\varphi} + \lambda_\omega\left(\boldsymbol{\varphi}^T\mathbf{M}\boldsymbol{\varphi} - 1\right). \tag{3.16}$$

The two adjoint equations are obtained by setting to zero the partial derivatives of $\mathcal{L}$ with respect to the state variables $\boldsymbol{\varphi}$ and $\omega$:

$$\frac{\partial\mathcal{L}}{\partial\boldsymbol{\varphi}} = \boldsymbol{\lambda}_\varphi^T\left(\mathbf{K} - \omega^2\mathbf{M}\right) + 2\lambda_\omega\boldsymbol{\varphi}^T\mathbf{M} = \mathbf{0}^T \tag{3.17a}$$

$$\frac{\partial \mathcal{L}}{\partial \omega} = 1 - 2\omega \boldsymbol{\lambda}_\varphi^T \mathbf{M}\boldsymbol{\varphi} = 0. \tag{3.17b}$$

To satisfy the first adjoint equation, we set $\lambda_\omega = 0$ and $\boldsymbol{\lambda}_\varphi = \beta\boldsymbol{\varphi}$, where $\beta$ is a constant. Using these results, the second adjoint equation becomes

$$1 - 2\omega\beta\boldsymbol{\varphi}^T \mathbf{M}\boldsymbol{\varphi} = 0. \tag{3.18}$$

Using the mass normalization condition (3.7), we find $\beta = 1/2\omega$. Finally, the sensitivity of $\omega$ is computed as the partial derivative of $\mathcal{L}$ with respect to $\mu$:

$$\frac{\mathrm{d}\omega}{\mathrm{d}\mu} = \frac{\partial \mathcal{L}}{\partial \mu} = \frac{\boldsymbol{\varphi}^T \left(\frac{\partial \mathbf{K}}{\partial \mu} - \omega^2 \frac{\partial \mathbf{M}}{\partial \mu}\right) \boldsymbol{\varphi}}{2\omega}. \tag{3.19}$$

### 3.2.3 System Derivatives

The last thing we need to address is the partial derivatives of the structural matrices **M** and **K** with respect to the design variables $\mu$. In general, the dependence of the system matrices on the chosen parameters can be complex and not easily expressed in closed form. In such cases, it is convenient to approximate the required partial derivatives using a *finite difference* scheme [6]. In this work, we adopt the forward difference method, which approximates the derivative of a matrix $\mathbf{A}(\mu)$ with respect to a parameter $\mu$ as

$$\frac{\partial \mathbf{A}(\mu)}{\partial \mu} \approx \frac{\mathbf{A}(\mu_0 + \Delta\mu) - \mathbf{A}(\mu_0)}{\Delta\mu}, \tag{3.20}$$

where $\mu_0$ is the nominal parameter value and $\Delta\mu$ is a small perturbation.

This approach requires $n_\mu + 1$ model evaluations, where $n_\mu$ denotes the number of design parameters. While conceptually simple and easy to implement, the method requires careful selection of $\Delta\mu$ to ensure numerical stability.

Moreover, since the number of model evaluations scales linearly with the number of parameters, the computational cost of this method can become significant in large-scale problems. Nonetheless, we adopt this approach in the example in Sect. 3.3, as the system's structure and dimensionality are tailored to keep the number of required evaluations manageable.

## 3.3 Optimization Results

The MEMS layout shown in Fig. 3.3 is discretized using 20 finite elements for each drive beam and five elements for each sense beam, with an additional two elements used for the connections between sense beams.

The device is fabricated from polysilicon [1], which is modeled as a linear, isotropic material. The mechanical properties used in the simulations are: Young's modulus $E = 161$ GPa, Poisson's ratio $\nu = 0.23$, and density $\rho = 2330\,\text{kg/m}^3$. The plane stress approximation is used, with a uniform out-of-plane thickness of 24 μm.

The design parameters $\boldsymbol{\mu}$ include the geometric dimensions of the primary structural components: the length and width of the drive beams ($L_d$ and $W_d$, respectively), the length and width of the folded sense beams ($L_s$ and $W_s$), and the width of the connections between the two folded sense beams ($W_c$). The lengths of the connecting beams $L_c$ are kept fixed to preserve the overall geometry of the proof mass and the drive frame. The nominal values of these parameters are summarized in Table 3.1.

In total, the design is governed by five parameters:

$$\boldsymbol{\mu} = \{L_d,\ L_s,\ W_d,\ W_s,\ W_c\}.$$

The target eigenfrequencies for the drive and sense modes are set to $\omega_d = 28$ kHz and $\omega_s = 26$ kHz, respectively. A spurious frequency $\omega_{\text{sp}}$ is also considered, defined as the lowest eigenfrequency in the spectrum, excluding $\omega_d$ and $\omega_s$.

To ensure the correct identification of the drive and sense modes during optimization, the Modal Assurance Criterion (MAC) [10] is used to track the mode shapes. This identification is crucial to prevent convergence issues resulting from mode switching. A graphical overview of the MAC is provided in Sect. 4.3.1.

Table 3.2 gives the initial values and the bounds for the parameters, expressed as deviations with respect to the nominal values of the initial layout (Table 3.1).

The MEMS gyroscope model, including the MPC-based formulation and the parametrized layout described above, is implemented using the open-source finite

**Table 3.1** Nominal parameters of the MEMS gyroscope layout in Fig. 3.3

| | $L_d$ | $L_s$ | $L_c$ | $W_d$ | $W_s$ | $W_c$ |
|---|---|---|---|---|---|---|
| Values [μm] | 165.0 | 120.0 | 10.4 | 4.00 | 3.40 | 15.0 |

**Table 3.2** Initial parameters, absolute bounds, and optimal values used in the MEMS optimization problem (3.1). All the values are expressed as deviations (e.g., $L_d = L_{d,0}+\Delta L_d$) with respect to their nominal values of the initial layout (Table 3.1)

| | $\Delta L_d$ | $\Delta L_s$ | $\Delta W_d$ | $\Delta W_s$ | $\Delta W_c$ |
|---|---|---|---|---|---|
| Initial values [μm] | 0.00 | 0.00 | 0.00 | 0.00 | 0.00 |
| Absolute bounds [μm] | 25.0 | 25.0 | 1.00 | 1.00 | 1.00 |
| Optimal values [μm] | −0.01 | 16.8 | 0.16 | 0.10 | −1.00 |

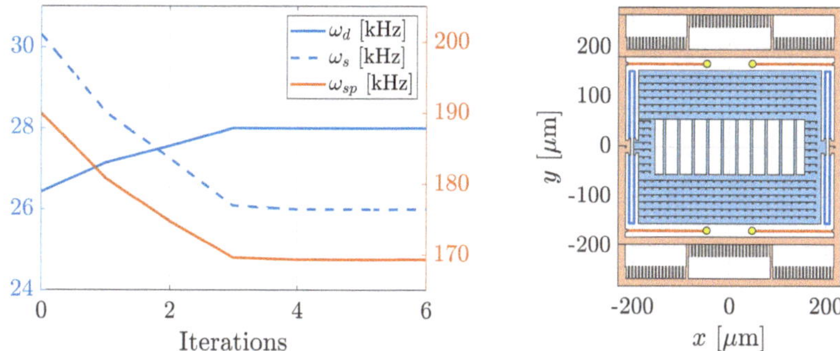


**Fig. 3.4** Optimization history (left) and optimal layout (right) of the MEMS optimization problem (3.1)

**Table 3.3** Characteristics of the drive comb-fingers electrodes of the MEMS gyroscope in [13]. The value of the alternate voltage corresponds to 12 dBm

| | Symbol | Value |
|---|---|---|
| Vacuum permittivity | $\epsilon_0$ | $8.854 \cdot 10^{-12}$ F/m |
| Thickness | $h$ | 24.0 µm |
| Number of comb-fingers | $N_f$ | 72 |
| Comb-fingers gap | $g$ | 1.90 µm |
| Rotor voltage | $v_{dc}$ | 15.0 V |
| Alternate peak voltage | $v_{ac}$ | 1.26 V |

element framework `YetAnotherFEcode` [9]. The optimization problem (3.1) is solved using MATLAB's gradient-based optimizer `fmincon`, configured with the `interior-point` algorithm.

The optimal layout, along with the optimization history, is shown in Fig. 3.4, while the optimal parameters are reported in Table 3.2. Although the optimization problem aims to maximize the spurious frequency $\omega_{sp}$, its value is reduced compared to the initial one to satisfy the constraints on the drive and sense frequencies.

### 3.3.1 Nonlinear Dynamic Response

The nonlinear dynamic response of the optimized MEMS gyroscope in Fig. 3.4 is evaluated using an $O(3)$ SSM as described in Chap. 2.

The characteristics of the drive comb-fingers electrodes [13], along with typical voltage settings, are given in Table 3.3.

Using these values, the actuation force is computed as [1]:

$$F_d = 2N_f\epsilon_0\frac{h}{g}v_{ac}v_{dc} \approx 0.30\,\mu\text{N}, \tag{3.21}$$

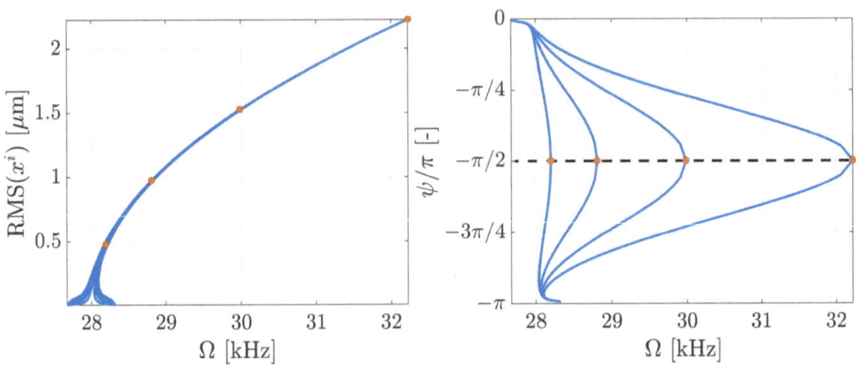


**Fig. 3.5** Frequency response amplitude (left) and phase (right) of the drive mode of the optimized MEMS gyroscope in Fig. 3.4. The peaks, marked with red dots, have phase $\psi = -\pi/2$

where the overall force is applied in the middle of the proof mass along the vertical direction to excite the drive mode of the gyroscope.

Using a Q-factor of 2200 [13], the frequency response associated with the drive mode is shown in Fig. 3.5 for different forcing levels up to the one computed from Eq. (3.21). The peaks of the frequency response curves, highlighted in red, are characterized by a phase $\psi = -\pi/2$.

Although the MEMS gyroscope was optimized in the linear regime for a target drive frequency of 28 kHz, the actual frequency at the nominal forcing level is significantly higher (Fig. 3.5). For example, at the maximum forcing level, the frequency response shows a peak at 32.2 kHz with an oscillation amplitude of 2.22 μm. If not considered, this nonlinear effect can alter the desired frequency mismatch between drive and sense modes, thus reducing the overall performance of the device. This effect highlights the importance of accounting for nonlinear dynamics in the optimization process, either to mitigate their impact or to strategically exploit them for enhanced robustness and performance [11, 15].

## References

1. Acar C, Shkel A (2009) MEMS vibratory gyroscopes. Springer, Berlin. https://doi.org/10.1007/978-0-387-09536-3
2. Bendsøe MP, Sigmund O (2004) Topology optimization: theory, methods, and applications. Springer, Berlin. https://doi.org/10.1007/978-3-662-05086-6
3. Benkhelifa E, Farnsworth M, Tiwari A, Bandi G, Zhu M (2010) Design and optimisation of microelectromechanical systems: a review of the state-of-the-art. Int J Des Eng 3:41. https://doi.org/10.1504/IJDE.2010.032822
4. Christensen PW, Klarbring A (2008) An introduction to structural optimization, vol 153, 1st edn. Springer, Dordrecht. https://doi.org/10.1007/978-1-4020-8666-3
5. Cook RD, Malkus DS, Plesha ME, Witt RJ (2001) Concepts and applications of finite element analysis, 4th edn. Wiley, Hoboken (2001)

6. Giannini D, Bonaccorsi G, Braghin F (2020) Size optimization of mems gyroscopes using substructuring. Eur J Mech A/Solids 84:104045. https://doi.org/10.1016/j.euromechsol.2020.104045
7. Giannini D, Bonaccorsi G, Braghin F (2021) Rapid prototyping of inertial mems devices through structural optimization. Sensors 21:5064. https://doi.org/10.3390/s21155064
8. Haronian D (1995) Maximizing microelectromechanical sensor and actuator sensitivity by optimizing geometry. Sens Actuat A Phys 50:223–236. https://doi.org/10.1016/0924-4247(95)01086-6
9. Jain S, Marconi J, Pozzi M, Tiso P (2026) YetAnotherFEcode (v1.5.0). Zenodo. https://doi.org/10.5281/zenodo.15257966
10. Kim TS, Kim YY (2000) Mac-based mode-tracking in structural topology optimization. Comput Struct 74:375–383. https://doi.org/10.1016/S0045-7949(99)00056-5
11. Li D, Shaw SW, Polunin PM (2022) Computational modeling of nonlinear dynamics and its utility in mems gyroscopes. J Struct Dyn. 1:217–235. https://doi.org/10.25518/2684-6500.96
12. Manzoni A, Quarteroni A, Salsa S (2021) Optimal control of partial differential equations, vol 207. Springer, Berlin. https://doi.org/10.1007/978-3-030-77226-0
13. Marconi J, Bonaccorsi G, Giannini D, Falorni L, Braghin F (2021) Exploiting nonlinearities for frequency-matched mems gyroscopes tuning. In: 2021 IEEE international symposium on inertial sensors and systems (INERTIAL). IEEE, Piscataway, pp 1–4 (2021). https://doi.org/10.1109/INERTIAL51137.2021.9430478
14. Pozzi M, Marconi J, Jain S, Li M, Braghin F (2025) Adjoint sensitivities for the optimization of nonlinear structural dynamics via spectral submanifolds. Proc R Soc A Math Phys Eng Sci 481:20250244. https://doi.org/10.1098/rspa.2025.0244
15. Rhoads JF, Shaw SW, Turner KL (2010) Nonlinear dynamics and its applications in micro- and nanoresonators. J Dyn Syst Measur Control 132:034001. https://doi.org/10.1115/1.4001333
16. Schiwietz D, Hörsting M, Weig EM, Degenfeld-Schonburg P, Wenzel M (2025) Shape optimization of eigenfrequencies in mems gyroscopes. Struct Multidiscip Optim 68:171. https://doi.org/10.1007/s00158-025-04112-w
17. Xia D, Kong L, Gao H (2015) Design and analysis of a novel fully decoupled tri-axis linear vibratory gyroscope with matched modes. Sensors 15:16929–16955. https://doi.org/10.3390/s150716929
18. Xia D, Kong L, Gao H (2015) A mode matched triaxial vibratory wheel gyroscope with fully decoupled structure. Sensors 15:28979–29002. https://doi.org/10.3390/s151128979

# Chapter 4
# Introduction to Topology Optimization

**Abstract** This chapter introduces topology optimization algorithms with a particular focus on density-based methods and their applications in linear structural dynamics.

## 4.1 Introduction

Topology optimization is a computational design method used to determine the optimal material layout within a given design space, based on specific performance criteria.

The foundation of topology optimization lies in the work of Bendsøe and Kikuchi [10] in the late 1980s, where they introduced the homogenization method for topology optimization of continuum structures. Their approach used a microstructural representation to find optimal material distributions that effectively balanced stiffness and weight, laying the groundwork for modern methods. Since this original work, different approaches to topology optimization have been proposed, including density methods [9, 11], evolutionary methods [51], the Level Set Method (LSM) [4, 5, 49], and, more recently, the method of Moving Morphable Component (MMC) [27]. A comprehensive review of topology optimization approaches can be found in [12, 43, 48].

Early applications of topology optimization focused on structural mechanics, aiming to improve stiffness, minimize weight, and control stress distribution within mechanical components [12]. These early approaches involved distributing a fixed amount of material across a design space to maximize stiffness or minimize compliance (the inverse of stiffness), thereby creating structures capable of withstanding loads with minimal deformation, a challenge known as the compliance minimization problem. Optimizing for stiffness was particularly valuable in fields where weight reduction was essential, such as aerospace, automotive, and civil engineering.

In the context of linear structural dynamics, the initial works of [21, 33, 34, 37] addressed the eigenfrequency maximization problem. Beyond maximizing fundamental eigenfrequencies, recent advancements have explored eigenfrequency gap maximization and targeted mode separation [23], which are particularly relevant

 
M. Pozzi, F. Braghin, *Structural Optimization for Nonlinear Dynamics*,
PoliMI SpringerBriefs, https://doi.org/10.1007/978-3-032-22600-6_4

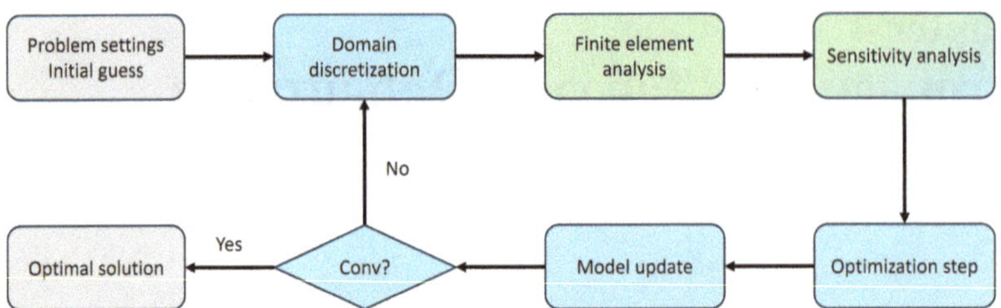


**Fig. 4.1** Topology optimization loop. Blue blocks represent elements related to the optimization approach, while green blocks are specific to the application. The sensitivity block is primarily application-dependent, with some influence from the optimization approach

for vibration isolation [17, 18, 26], phononic crystals [20, 42], and MEMS resonators [13, 24, 25]. Further advancements explored robustness aspects, including manufacturing uncertainties [32, 39] and temperature uncertainties [38]. Parallel to eigenfrequency optimization, several works have been devoted to optimizing frequency response curves [3, 35, 40], addressing additional dynamic performance criteria [45, 46].

Despite the specific application, most topology optimization approaches share a similar optimization architecture [6]. The purpose of this chapter is to describe the standard building blocks used in a topology optimization loop (Fig. 4.1), focusing on the density-based approach for eigenfrequency optimization. Benchmark cases are used to illustrate the steps: the compliance minimization of a Messerschmitt-Bölkow-Blohm (MBB) beam (a classic example in topology optimization), the frequency maximization of a mechanical resonator, and the frequency assignment of a mechanical resonator.

## 4.2 Domain Discretization

The first step in a topology optimization loop involves discretizing the design domain. For density-based methods, a structured grid of square elements is typically used. This grid serves as the finite element mesh for the analysis, providing a basis for representing material distribution. Each element is assigned a density value $\mu$ between 0 and 1: a density of 1 represents material (solid), 0 represents void (no material), and values between 0 and 1 indicate regions partially intersected by the structure's boundary (Fig. 4.2). Typically, element densities are denoted by the symbol $\rho$. However, to distinguish them from material density, we use the symbol $\mu$ for element densities, which serve as the design variables in the optimization problem.

All the densities $\mu$ are collected in the vector $\boldsymbol{\mu} \in \mathbb{R}^{N_{\mathrm{el}}}$, where $N_{\mathrm{el}}$ is the total number of elements. Typically, the number of elements goes from a few thousand for small 2D problems up to billions for large-scale 3D problems [1, 8].

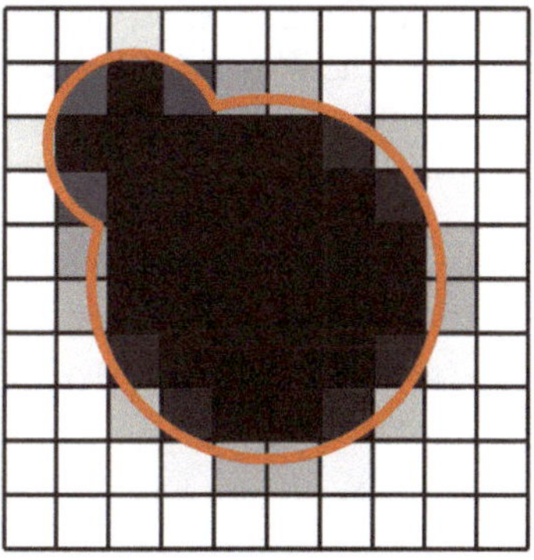

**Fig. 4.2** Example of structure (solid red line) discretized using the density method

**Fig. 4.3** Example of the regularization filter (4.1) with $R_{\text{filt}} = 4$ elements. The right image is obtained by applying the filter to the left one

The densities are used to interpolate the material properties during the finite element assembly of the structural matrices. However, the direct use of $\mu$ leads to mesh-dependent solutions, where the optimization might introduce small features with a size comparable to that of an element of the grid. If not properly addressed, this dependence on the mesh affects the scale of the design [50]. Moreover, the use of a coarse discretization might lead to checkerboard patterns [22, 29]. To obtain a well-posed optimization problem and to avoid mesh-dependent solutions, various techniques have been proposed [12, 44]. A popular choice is the convolution filter proposed by Bourdin [14] and Bruns and Tortorelli [15]. The idea is to define a filtered density $\tilde{\mu}$ as the weighted average of densities $\mu$ of the neighbor elements (Fig. 4.3). The filter is defined as

$$\tilde{\mu}_e = \frac{\sum_{j \in N_e} w_{e,j} \mu_j}{\sum_{j \in N_e} w_{e,j}}, \tag{4.1}$$

where $\mathcal{N}_e$ is the set of neighbor elements that lie within a radius $R_{\text{filt}}$ around the $e$-th element. The weight $w_{e,j}$ is computed as

$$w_{e,j} = R_{\text{filt}} - |\boldsymbol{x}_j - \boldsymbol{x}_e|, \tag{4.2}$$

where $\boldsymbol{x}_j$ and $\boldsymbol{x}_e$ contain the coordinates of the center of the elements.

The filter (4.1) can be implemented efficiently using a sparse matrix representation as done by Andreassen et al. [7]. In this case, the vector of filtered densities $\tilde{\boldsymbol{\mu}}$ is computed as

$$\tilde{\boldsymbol{\mu}} = \mathbf{H}\boldsymbol{\mu}, \tag{4.3}$$

where $\mathbf{H}$ is the filter matrix that contains the weights computed using Eq. (4.2).

As previously said, the addition of the filter prevents checkerboard problems and mesh-dependent solutions. Additionally, the radius $R_{\text{filt}}$ provides length scale control during optimization, thereby limiting the minimum feature size of the optimal design. However, this regularization filter may lead to non-physical regions, known as gray transition regions, with an intermediate density value between 0 and 1. To obtain a well-defined design, projection techniques are used to project the filtered densities into the 0/1 space. There exist many projection functions in the literature, but in this thesis, the projection proposed by Wang et al. [50] is used:

$$\bar{\tilde{\mu}}_e = \frac{\tanh(\beta\eta) + \tanh(\beta(\tilde{\mu}_e - \eta))}{\tanh(\beta\eta) + \tanh(\beta(1 - \eta))}, \tag{4.4}$$

where $\bar{\tilde{\mu}}_e$ is the projected density of the $e$-th element. The parameter $\beta$ controls the slope of the transition between 0 and 1. The parameter $\eta$ controls the projection threshold: all the density values above $\eta$ are projected toward 1, while all the values below $\eta$ are projected toward 0 (Fig. 4.4).

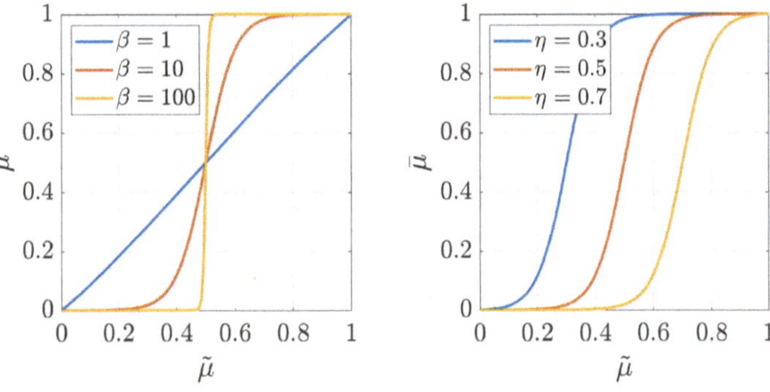


**Fig. 4.4** Example of the projection function (4.4)

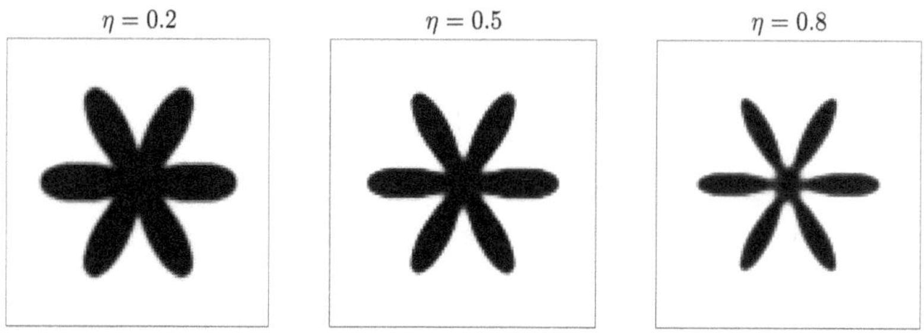


**Fig. 4.5** Effect of the projection function (4.4) on a star shape. The nominal shape (middle) is obtained using $\eta = 0.5$. The under-etched (left) and over-etched (right) versions are obtained using $\eta = 0.2$ and $\eta = 0.8$, respectively

The nominal value for $\eta$ is 0.5. Still, there are cases in which the projection threshold is either set to higher or lower values to obtain under-etched or over-etched versions of the nominal structure (Fig. 4.5). These projections are used to include manufacturing uncertainties in the topology optimization problem [41].

The projected densities are finally used to obtain the physical densities $\hat{\mu}$ that are used in the finite element assembly. According to the Solid Isotropic Material with Penalization (SIMP) scheme [11], the physical densities are defined as

$$\hat{\mu}_e = \mu_0 + (1 - \mu_0)\bar{\mu}_e^p, \tag{4.5}$$

where $p$ is the penalization power used to further penalize the gray transition regions, and $\mu_0$ is an arbitrarily small density of the void element used to avoid singularities in the structural matrices. Typically, $p = 3$ and $\mu_0 = 10^{-6}$ are used. In some problems, the penalization power $p$ is gradually increased during the optimization, starting from 1.

## 4.3 Finite Element Analysis

The physical densities $\hat{\mu}$ are used in the finite element assembly of the structural matrices. For example, the stiffness matrix is computed as

$$\mathbf{K} = \bigcup_{e=1}^{N_{\text{el}}} \hat{\mu}_e \mathbf{K}_e, \tag{4.6}$$

where $\mathbf{K}_e$ is the stiffness matrix of element $e$, and the operator $\bigcup$ represents the finite element assembly across all the elements of the mesh. Since a structured grid is used, the element matrices are the same for all elements. Therefore, they can be computed only once, thus reducing the overall computational cost.

Similarly, the mass matrix is assembled as

$$\mathbf{M} = \bigcup_{e=1}^{N_{\text{el}}} \hat{\mu}_e \mathbf{M}_e, \tag{4.7}$$

where $\mathbf{M}_e$ is the mass matrix of element $e$.

**Finite Element Library**

The element stiffness matrix is computed as

$$\mathbf{K}_e = \int_{\Omega_e} \mathbf{B}^T \mathbf{C}_e \mathbf{B} \mathrm{d}x,$$

where $\Omega_e$ is the domain of the finite element, $\mathbf{C}_e$ is the elasticity matrix, and $\mathbf{B}$ is the strain-displacement matrix.

Similarly, the element mass matrix is computed as

$$\mathbf{M}_e = \int_{\Omega_e} \rho \mathbf{N}^T \mathbf{N} \mathrm{d}x,$$

where $\rho$ is the material density, and $\mathbf{N}$ is the matrix of shape functions.

After the assembly, the finite element analysis is used to solve the physics of the problem. For example, to evaluate the compliance, the stationary problem is solved for the displacement vector $\boldsymbol{u}$:

$$\mathbf{K}\boldsymbol{u} = \boldsymbol{f}, \tag{4.8}$$

where $\boldsymbol{f}$ is the vector of external loads that do not depend on the densities. Using the vector $\boldsymbol{u}$, the compliance is evaluated as $C = \boldsymbol{f}^T \boldsymbol{u}$.

The eigenfrequencies $\omega_i$, and the corresponding mode shapes $\boldsymbol{\varphi}_i$, are computed by solving the generalized eigenvalue problem:

$$(-\omega_i^2 \mathbf{M} + \mathbf{K})\boldsymbol{\varphi}_i = \mathbf{0} \tag{4.9}$$

Typically, in topology optimization, the mode shapes are mass-normalized to simplify the sensitivity analysis (Sect. 4.4):

$$\boldsymbol{\varphi}_i^T \mathbf{M} \boldsymbol{\varphi}_i = 1 \tag{4.10a}$$

$$\boldsymbol{\varphi}_i^T \mathbf{K} \boldsymbol{\varphi}_i = \omega_i^2. \tag{4.10b}$$

### 4.3.1 *Mode Switching*

When dealing with optimization problems involving eigenfrequencies, it is essential to track the desired modes [13]. As the optimization advances, a change in the structure's layout may alter the order of the modes. This effect is known as *mode switching* [33–35]. Suppose the objective function or constraints are defined based on a fixed mode order. In that case, their values may experience sudden jumps when the modes switch positions, leading to convergence problems or unpredictable results. Therefore, it is important to address the mode switching issue: to make sure the correct mode is correctly tracked, a practical approach is to use the Modal Assurance Criterion (MAC) [30], which provides a measure of the similarities between two mode shapes:

$$\mathrm{MAC}_{ij} = \frac{(\boldsymbol{\varphi}_i^T \boldsymbol{\varphi}_j)^2}{(\boldsymbol{\varphi}_i^T \boldsymbol{\varphi}_i)(\boldsymbol{\varphi}_j^T \boldsymbol{\varphi}_j)}. \tag{4.11}$$

At the beginning of the optimization, the shape of the target mode is identified. Then, at each iteration, a large enough set of modes is computed. These modes are compared to the target one by evaluating the MAC value, and the mode shape with the highest MAC is the target one. The MAC is also applicable to discard localized modes [19, 31] associated with gray areas or floating material (a small portion of material detached from the main structure).

## 4.4 Sensitivity Analysis

In every gradient-based optimization, including topology optimization, the sensitivity analysis is perhaps the most important part, as it drives the entire optimization process. For example, consider the compliance $C = \boldsymbol{f}^T \boldsymbol{u}$, which depends on the design variables $\boldsymbol{\mu}$ through all the filtering stages (regularization filter, projection function, and SIMP scheme). The sensitivity with respect to the density $\mu_e$ of the $e$-th element is derived using the chain rule:

$$\frac{\mathrm{d}C}{\mathrm{d}\mu_e} = \sum_{j=1}^{N_{\mathrm{el}}} \frac{\mathrm{d}C}{\mathrm{d}\hat{\mu}_j} \frac{\mathrm{d}\hat{\mu}_j}{\mathrm{d}\bar{\mu}_j} \frac{\mathrm{d}\bar{\mu}_j}{\mathrm{d}\tilde{\mu}_j} \frac{\mathrm{d}\tilde{\mu}_j}{\mathrm{d}\mu_e}. \tag{4.12}$$

This expression involves the derivative of the regularization filter, the projection function, and the SIMP scheme:

$$\frac{\mathrm{d}\tilde{\mu}_j}{\mathrm{d}\mu_e} = \begin{cases} \frac{w_{j,e}}{\sum_{l \in N_j} w_{j,l}} & \text{if } e \in N_j \\ 0 & \text{otherwise} \end{cases} \tag{4.13}$$

$$\frac{\mathrm{d}\bar{\mu}_j}{\mathrm{d}\tilde{\mu}_j} = \beta \frac{1 - \tanh^2\left(\beta(\tilde{\mu}_j - \eta)\right)}{\tanh\left(\beta\eta\right) + \tanh\left(\beta(1-\eta)\right)} \tag{4.14}$$

$$\frac{\mathrm{d}\hat{\mu}_j}{\mathrm{d}\bar{\mu}_j} = p(1 - \hat{\mu}_0)\bar{\mu}_j^{p-1}. \tag{4.15}$$

Equations (4.13), (4.14), and (4.15) do not depend on the particular optimization problem and can be generally used.

To obtain the derivative of the compliance $C$ with respect to $\hat{\mu}_j$, we use the *adjoint method* [36] as described in Sect. 3.2.1. To do so, we consider $\hat{\mu}_j$ as the *control variable* and $\boldsymbol{u} \in \mathbb{R}^n$ as the *state variable*. The latter is obtained by solving the *state equation* which, in this case, is the stationary equation (4.8). Then, the *Lagrangian function* $\mathcal{L}$ is defined as

$$\mathcal{L}(\hat{\mu}_j, \boldsymbol{u}, \boldsymbol{\lambda}) = C + \boldsymbol{\lambda}^T(\mathbf{K}\boldsymbol{u} - \boldsymbol{f}), \tag{4.16}$$

where $\boldsymbol{\lambda} \in \mathbb{R}^n$ is the *adjoint variable*.

The *adjoint equation* is then obtained by setting to zero the partial derivative of $\mathcal{L}$ with respect to $\boldsymbol{u}$:

$$\frac{\partial \mathcal{L}}{\partial \boldsymbol{u}} = \boldsymbol{f}^T + \boldsymbol{\lambda}^T\mathbf{K} = \mathbf{0}^T. \tag{4.17}$$

Equation (4.17) is a linear system of size $n \times n$ whose solution is $\boldsymbol{\lambda}$. Notice that, for the compliance example, the solution of the adjoint equation is $\boldsymbol{\lambda} = -\boldsymbol{u}$. In this case, the problem is called *self-adjoint*, meaning that there is no need to solve the adjoint equation because the adjoint variable is already known.

After computing the adjoint variable, the sensitivity of $C$ is the partial derivative of $\mathcal{L}$ with respect to $\mu$:

$$\frac{\mathrm{d}C}{\mathrm{d}\hat{\mu}_j} = \frac{\partial \mathcal{L}}{\partial \hat{\mu}_j} = -\boldsymbol{u}^T \frac{\partial \mathbf{K}}{\partial \hat{\mu}_j}\boldsymbol{u}, \tag{4.18}$$

where the derivative of $\mathbf{K}$ with respect to $\hat{\mu}_j$ is simply the element stiffness matrix $\mathbf{K}_j$. Therefore, in the product $\boldsymbol{u}^T \frac{\partial \mathbf{K}}{\partial \hat{\mu}_j}\boldsymbol{u}$, only the displacements relative to the nodes of the $j$-th element are involved. By defining these displacements $\boldsymbol{u}_j$, the sensitivity of the compliance becomes

$$\frac{\mathrm{d}C}{\mathrm{d}\hat{\mu}_j} = -\boldsymbol{u}_j^T\mathbf{K}_j\boldsymbol{u}_j. \tag{4.19}$$

The adjoint formulation of the sensitivity requires the solution of at most one adjoint system for each state variable. Therefore, the computational cost does not depend on the number of design variables, making this approach particularly suitable for topology optimization, where even small problems have thousands of design variables.

The adjoint method is also used for eigenvalue problems. Here we report the final expression for the adjoint sensitivity of a generic eigenfrequency $\omega$, which was derived in Sect. 3.2.2. Provided that the mode shapes are mass normalized, the adjoint sensitivity of $\omega$ with respect to $\hat{\mu}_j$ is

$$\frac{d\omega}{d\hat{\mu}_j} = \frac{1}{2\omega}\boldsymbol{\varphi}^T\left(-\omega^2\frac{\partial \mathbf{M}}{\partial \hat{\mu}_j} + \frac{\partial \mathbf{K}}{\partial \hat{\mu}_j}\right)\boldsymbol{\varphi}. \tag{4.20}$$

where the derivatives of $\mathbf{K}$ and $\mathbf{M}$ with respect to $\hat{\mu}_j$ are the element matrices $\mathbf{K}_j$ and $\mathbf{M}_j$. Similarly to the compliance case, the sensitivity is evaluated at the element level:

$$\frac{d\omega}{d\hat{\mu}_j} = \frac{1}{2\omega}\boldsymbol{\varphi}_j^T\left(-\omega^2\mathbf{M}_j + \mathbf{K}_j\right)\boldsymbol{\varphi}_j. \tag{4.21}$$

## 4.5 Optimization Step

Developed by Svanberg [47], the Method of Moving Asymptotes (MMA) is a widely used optimization algorithm in structural optimization, particularly in topology optimization. It is an iterative method that formulates the optimization problem as a sequence of subproblems. Each subproblem is a convex approximation of the original nonlinear problem, simplified by linear and asymptotic terms that "move" from iteration to iteration. At each iteration, the sensitivities are used to formulate the subproblem. Then the MMA is used to update the design variables of the optimization problem, i.e., the densities of the elements.

After the update, the algorithm checks for convergence. Typically, this is determined based on a set of criteria, which can include:

- *Change in objective function*: convergence is often declared if the relative change in the objective function falls below a predefined tolerance, typically around 0.1% or less between consecutive iterations.
- *Change in design variables*: the change in design variables (the element densities) should fall below a small tolerance threshold, indicating that the structure is no longer significantly evolving.
- *Residual in constraints*: for constrained problems, the residuals of these constraints should remain within an acceptable range across iterations.

In practice, multiple convergence criteria are often combined to ensure robust and reliable optimization, particularly for complex designs with nonlinear objectives and multiple constraints. Moreover, as a safety measure, a maximum iteration count is set to avoid excessive computational time if other criteria are not met due to oscillatory behavior or slow convergence.

Another important aspect is the maximum allowable change in the densities ($\Delta\mu$) at each iteration. Setting this value too low can slow convergence, while setting it too high may cause convergence issues. The optimal value depends on the specific application; based on the author's experience, a good choice is between 0.01 and 0.1.

## 4.6 Numerical Examples

In this section, three numerical examples are presented: the compliance minimization of an MBB beam, the frequency maximization of a mechanical resonator, and the frequency assignment of a mechanical resonator.

For both examples, consider a 2D rectangular design domain of size $1000 \times 500\,\mu m$ discretized using a regular grid with $200 \times 100$ square elements with linear shape functions. The plane-stress approximation is used, with a uniform thickness of $24\,\mu m$. The material is polysilicon [2], with Young's modulus $E = 161\,GPa$, Poisson's ratio $\nu = 0.23$, and density $\rho = 2330\,kg/m^3$.

A regularization filter with radius $R_{filt} = 4$ is used, along with a projection function with $\beta = 10$ and $\eta = 0.5$. The SIMP scheme is used to interpolate both the stiffness and mass matrices, with $\mu_0 = 10^{-6}$ and $p = 3$.

The limit on the maximum density change is set to $\Delta\mu = 0.1$.

The optimization problems are implemented in `YetAnotherFEcode` [28].

### 4.6.1 *Compliance Minimization of an MBB Beam*

The scheme of the MBB beam is shown in Fig. 4.6: the left side is clamped, while the right one has rollers that prevent horizontal displacement. An external load of magnitude $f = 10^3\,\mu N$ is applied downwards at the top-right corner of the domain.

**Fig. 4.6** Problem settings and boundary conditions of the MBB beam

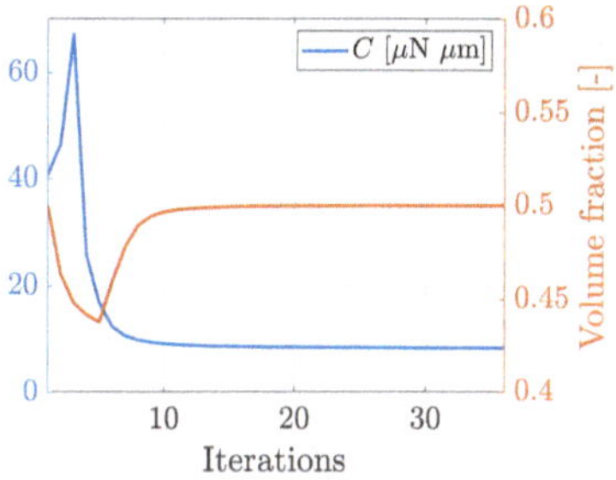

**Fig. 4.7** Optimization history (left) and optimal layout (right) of the compliance minimization problem (4.22)

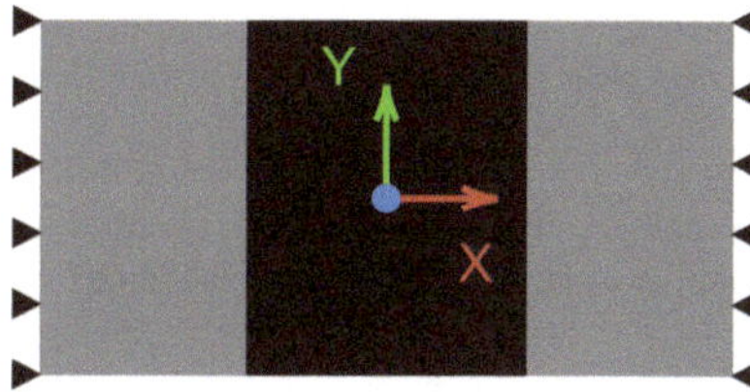


**Fig. 4.8** Problem settings and boundary conditions of a single-mass mechanical resonator

The compliance minimization problem is formulated as

$$\begin{aligned} \min_{\boldsymbol{\mu}} \quad & \boldsymbol{f}^T \boldsymbol{u} \\ \text{s.t.} \quad & V - V_{\max} \leq 0 \\ & \mathbf{0} \leq \boldsymbol{\mu} \leq \mathbf{1}, \end{aligned} \tag{4.22}$$

where $V$ is the total volume of the structure with upper limit $V_{\max} = 50\%$ of the design domain.

The optimization problem reached convergence in 37 iterations (Fig. 4.7), with the optimal layout showing a compliance $C = 8.48\,\mu\text{N}\mu\text{m}$.

### *4.6.2 Frequency Maximization of a Resonator*

The scheme of the single-mass mechanical resonator is shown in Fig. 4.8: the left and right sides are clamped. The proof mass (black region) occupies the 40% of the design domain, and remains solid ($\mu_e = 1$) for the entire optimization.

The optimization aims at maximizing the frequency $\omega_y$, which corresponds to the mode shape with the maximum displacement of the proof mass along the $y$ direction (Fig. 4.8). At each iteration, the MAC is used to track the target mode and to discard localized ones (Fig. 4.9).

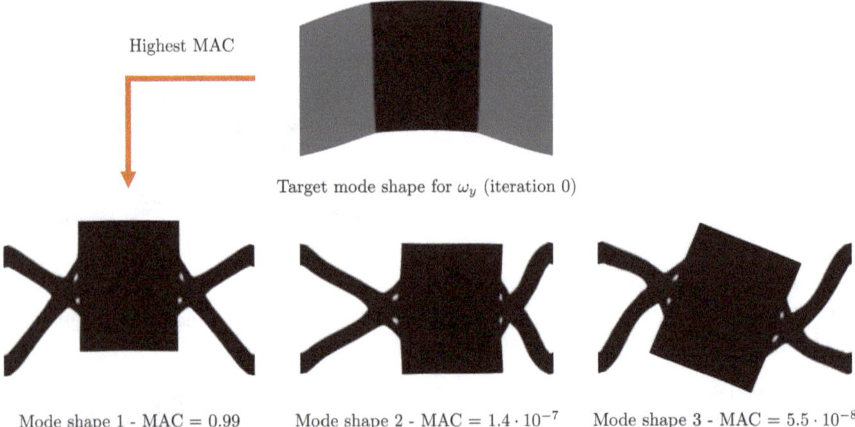


**Fig. 4.9** Graphical representation of the MAC used to track the mode shape corresponding to $\omega_Y$: at each iteration, a set of modes (bottom) are compared to the reference one (top) identified at the first iteration. The target mode is the one with the highest MAC

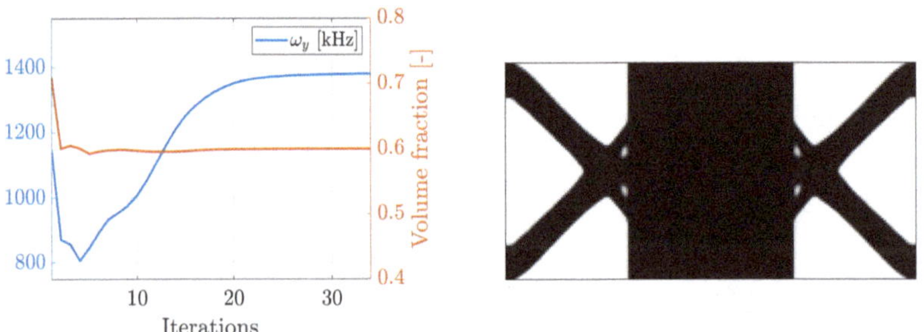


**Fig. 4.10** Optimization history (left) and optimal layout (right) of the frequency maximization problem (4.23)

The problem is formulated as

$$
\begin{aligned}
\max_{\boldsymbol{\mu}} \quad & \omega_y \\
\text{s.t.} \quad & V - V_{\max} \leq 0 \\
& \mathbf{0} \leq \boldsymbol{\mu} \leq \mathbf{1},
\end{aligned}
\tag{4.23}
$$

where $V$ is the total volume of the structure with upper limit $V_{\max} = 60\%$ of the design domain.

The optimization problem reached convergence in 35 iterations (Fig. 4.10), with the optimal layout showing a frequency $\omega_y = 1.38 \cdot 10^3$ kHz.

### 4.6.3 Frequency Assignment of a Resonator

Let us consider the same scheme for the single-mass mechanical resonator in Fig. 4.8. We try now to impose a target value on the eigenfrequency $\omega_y$. This problem is particularly relevant to the design of MEMS resonators.

To avoid connectivity issues and material detachment problems [13, 16], we set up the optimization problem to maximize the eigenfrequency $\omega_x$, which corresponds to the mode shape with the maximum displacement of the proof mass along the $x$ direction (Fig. 4.8). As for the previous example, the MAC is used to track both the target modes corresponding to $\omega_x$ and $\omega_y$.

The problem is formulated as

$$\begin{aligned} \max_{\mu} \quad & \omega_x \\ \text{s.t.} \quad & \left(\omega_y/\omega_{y,\text{target}} - 1\right)^2 - \epsilon^2 \leq 0 \\ & V - V_{\max} \leq 0 \\ & \mathbf{0} \leq \boldsymbol{\mu} \leq \mathbf{1}, \end{aligned} \tag{4.24}$$

where $V$ is the total volume of the structure with upper limit $V_{\max} = 60\%$ of the design domain, and $\epsilon = 10^{-3}$ is to impose the frequency $\omega_y$ to its target value $\omega_{y,\text{target}} = 1000$ kHz within a tolerance of 0.1%.

The optimization problem reached convergence in 51 iterations (Fig. 4.11), with the optimal layout showing frequencies $\omega_y = 999.1$ kHz and $\omega_x = 2.49 \cdot 10^3$ kHz.

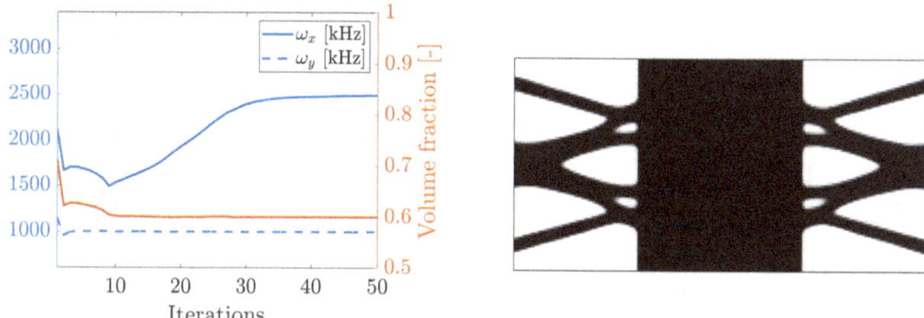


**Fig. 4.11** Optimization history (left) and optimal layout (right) of the frequency assignment problem (4.24)

## References


1. Aage N, Andreassen E, Lazarov BS, Sigmund O (2017) Giga-voxel computational morphogenesis for structural design. Nature 550:84–86. https://doi.org/10.1038/nature23911
2. Acar C, Shkel A (2009) MEMS vibratory gyroscopes. Springer, Berlin. https://doi.org/10.1007/978-0-387-09536-3
3. Allaire G, Jouve F (2005) A level-set method for vibration and multiple loads structural optimization. Comput Methods Appl Mech Eng 194:3269–3290. https://doi.org/10.1016/j.cma.2004.12.018
4. Allaire G, Jouve F, Toader AM (2002) A level-set method for shape optimization. Comptes Rendus Math 334:1125–1130. https://doi.org/10.1016/S1631-073X(02)02412-3
5. Allaire G, Jouve F, Toader AM (2004) Structural optimization using sensitivity analysis and a level-set method. J Comput Phys 194:363–393. https://doi.org/10.1016/J.JCP.2003.09.032
6. Andreasen CS, Elingaard MO, Aage N (2020) Level set topology and shape optimization by density methods using cut elements with length scale control. Struct Multidiscip Optim 62:685–707. https://doi.org/10.1007/s00158-020-02527-1
7. Andreassen E, Clausen A, Schevenels M, Lazarov BS, Sigmund O (2011) Efficient topology optimization in matlab using 88 lines of code. Struct Multidiscip Optim 43:1–16. https://doi.org/10.1007/s00158-010-0594-7
8. Baandrup M, Sigmund O, Polk H, Aage N (2020) Closing the gap towards super-long suspension bridges using computational morphogenesis. Nat Commun 11:2735. https://doi.org/10.1038/s41467-020-16599-6
9. Bendsøe MP (1989) Optimal shape design as a material distribution problem. Struct Optim 1:193–202. https://doi.org/10.1007/BF01650949
10. Bendsøe MP, Kikuchi N (1988) Generating optimal topologies in structural design using a homogenization method. Comput Methods Appl Mech Eng 71:197–224. https://doi.org/10.1016/0045-7825(88)90086-2
11. Bendsøe MP, Sigmund O (1999) Material interpolation schemes in topology optimization. Arch Appl Mech 69:635–654. https://doi.org/10.1007/S004190050248
12. Bendsøe MP, Sigmund O (2004) Topology optimization: theory, methods, and applications. Springer, Berlin. https://doi.org/10.1007/978-3-662-05086-6
13. Bonaccorsi G, Pozzi M, Hyun J, Kim HA, Braghin F (2025) Connectivity constraints for eigenvalue reduction in level-set topology optimization. Comput Struct 316:107865. https://doi.org/10.1016/j.compstruc.2025.107865
14. Bourdin B (2001) Filters in topology optimization. Int J Numer Methods Eng 50:2143–2158. https://doi.org/10.1002/nme.116
15. Bruns TE, Tortorelli DA (2001) Topology optimization of non-linear elastic structures and compliant mechanisms. Comput Methods Appl Mech Eng 190:3443–3459. https://doi.org/10.1016/S0045-7825(00)00278-4
16. Cool V, Aage N, Sigmund O (2025) A practical review on promoting connectivity in topology optimization. Struct Multidiscip Optim 68:73. https://doi.org/10.1007/s00158-025-04004-z
17. Cool V, Sigmund O, Aage N, Naets F, Deckers E (2024) Vibroacoustic topology optimization for sound transmission minimization through sandwich structures. J Sound Vib 568:117959. https://doi.org/10.1016/j.jsv.2023.117959
18. Cool V, Sigmund O, Aage N (2025) Metamaterial design with vibroacoustic bandgaps through topology optimization. Comput Methods Appl Mech Eng 436:117744. https://doi.org/10.1016/j.cma.2025.117744
19. Dalklint A, Wallin M, Tortorelli DA (2021) Structural stability and artificial buckling modes in topology optimization. Struct Multidiscip Optim 64:1751–1763. https://doi.org/10.1007/s00158-021-03012-z
20. Dalklint A, Wallin M, Bertoldi K, Tortorelli D (2022) Tunable phononic bandgap materials designed via topology optimization. J Mech Phys Solids 163:104849. https://doi.org/10.1016/j.jmps.2022.104849

21. Díaaz AR, Kikuchi N (1992) Solutions to shape and topology eigenvalue optimization problems using a homogenization method. Int J Numer Methods Eng 35:1487–1502. https://doi.org/10.1002/nme.1620350707
22. Díaz A, Sigmund O (1995) Checkerboard patterns in layout optimization. Struct Optim 10:40–45. https://doi.org/10.1007/BF01743693
23. Du J, Olhoff N (2007) Topological design of freely vibrating continuum structures for maximum values of simple and multiple eigenfrequencies and frequency gaps. Struct Multidiscip Optim 34:91–110. https://doi.org/10.1007/s00158-007-0101-y
24. Giannini D, Braghin F, Aage N (2020) Topology optimization of 2d in-plane single mass mems gyroscopes. Struct Multidiscip Optim 62:2069–2089. https://doi.org/10.1007/s00158-020-02595-3
25. Giannini D, Aage N, Braghin F (2022) Topology optimization of mems resonators with target eigenfrequencies and modes. Eur J Mech A/Solids 91:104352. https://doi.org/10.1016/j.euromechsol.2021.104352
26. Giannini D, Schevenels M, Reynders EP (2025) Topology optimization design of multi-modal resonators for metamaterial panels with maximized broadband vibroacoustic attenuation. J Sound Vib 595:118691. https://doi.org/10.1016/j.jsv.2024.118691
27. Guo X, Zhang W, Zhong W (2014) Doing topology optimization explicitly and geometrically - a new moving morphable components based framework. J Appl Mech 81. https://doi.org/10.1115/1.4027609
28. Jain S, Marconi J, Pozzi M, Tiso P (2026) YetAnotherFEcode (v1.5.0). Zenodo. https://doi.org/10.5281/zenodo.15257966
29. Jog CS, Haber RB (1996) Stability of finite element models for distributed-parameter optimization and topology design. Comput Methods Appl Mech Eng **130**:203–226. https://doi.org/10.1016/0045-7825(95)00928-0
30. Kim TS, Kim YY (2000) Mac-based mode-tracking in structural topology optimization. Comput Struct 74:375–383. https://doi.org/10.1016/S0045-7949(99)00056-5
31. Li Z, Shi T, Xia Q (2017) Eliminate localized eigenmodes in level set based topology optimization for the maximization of the first eigenfrequency of vibration. Adv Eng Softw 107:59–70. https://doi.org/10.1016/j.advengsoft.2016.12.001
32. Li Z, Shi T, Xia L, Xia Q (2019) Maximizing the first eigenfrequency of structures subjected to uniform boundary erosion through the level set method. Eng Comput 35:21–33. https://doi.org/10.1007/s00366-018-0580-z
33. Ma ZD, Cheng HC, Kikuchi N (1994) Structural design for obtaining desired eigenfrequencies by using the topology and shape optimization method. Comput Syst Eng 5:77–89. https://doi.org/10.1016/0956-0521(94)90039-6
34. Ma ZD, Kikuchi N, Cheng HC (1995) Topological design for vibrating structures. Comput Methods Appl Mech Eng 121:259–280. https://doi.org/10.1016/0045-7825(94)00714-X
35. Ma ZD, Kikuchi N, Hagiwara I (1993) Structural topology and shape optimization for a frequency response problem. Comput Mech 13:157–174. https://doi.org/10.1007/BF00370133
36. Manzoni A, Quarteroni A, Salsa S (2021) Optimal control of partial differential equations, vol 207. Springer, Berlin. https://doi.org/10.1007/978-3-030-77226-0
37. Pedersen N (2000) Maximization of eigenvalues using topology optimization. Struct Multidiscip Optim 20:2–11. https://doi.org/10.1007/s001580050130
38. Pozzi M, Bonaccorsi G, Braghin F (2023) A temperature-robust level-set approach for eigenfrequency optimization. Struct Multidiscip Optim 66:173. https://doi.org/10.1007/s00158-023-03622-9
39. Pozzi M, Bonaccorsi G, Kim HA, Braghin F (2023) Robust structural optimization in presence of manufacturing uncertainties through a boundary-perturbation method. Struct Multidiscip Optim 66:120. https://doi.org/10.1007/s00158-023-03573-1
40. Shu L, Wang MY, Fang Z, Ma Z, Wei P (2011) Level set based structural topology optimization for minimizing frequency response. J Sound Vib 330:5820–5834. https://doi.org/10.1016/j.jsv.2011.07.026

41. Sigmund O (2009) Manufacturing tolerant topology optimization. Acta Mech Sin 25:227–239. https://doi.org/10.1007/s10409-009-0240-z
42. Sigmund O, Jensen JS (2003) Systematic design of phononic bandgap materials and structures by topology optimization. Philos Trans R Soc Lond A Math Phys Eng Sci 361:1001–1019. https://doi.org/10.1098/rsta.2003.1177
43. Sigmund O, Maute K (2013) Topology optimization approaches: a comparative review. Struct Multidiscip Optim 48:1031–1055. https://doi.org/10.1007/s00158-013-0978-6
44. Sigmund O, Petersson J (1998) Numerical instabilities in topology optimization: a survey on procedures dealing with checkerboards, mesh-dependencies and local minima. Struct Optim 16:68–75. https://doi.org/10.1007/BF01214002
45. Silva OM, Neves MM, Lenzi A (2019) A critical analysis of using the dynamic compliance as objective function in topology optimization of one-material structures considering steady-state forced vibration problems. J Sound Vib 444:1–20. https://doi.org/10.1016/j.jsv.2018.12.030
46. Silva OM, Neves MM, Lenzi A (2020) On the use of active and reactive input power in topology optimization of one-material structures considering steady-state forced vibration problems. J Sound Vib 464:114989. https://doi.org/10.1016/j.jsv.2019.114989
47. Svanberg K (1987) The method of moving asymptotes - a new method for structural optimization. Int J Numer Methods Eng 24:359–373. https://doi.org/10.1002/nme.1620240207
48. van Dijk NP, Maute K, Langelaar M, van Keulen F (2013) Level-set methods for structural topology optimization: a review. Struct Multidiscip Optim 48:437–472. https://doi.org/10.1007/s00158-013-0912-y
49. Wang MY, Wang X, Guo D (2003) A level set method for structural topology optimization. Comput Methods Appl Mech Eng 192:227–246. https://doi.org/10.1016/S0045-7825(02)00559-5
50. Wang F, Lazarov BS, Sigmund O (2011) On projection methods, convergence and robust formulations in topology optimization. Struct Multidiscip Optim 43:767–784. https://doi.org/10.1007/s00158-010-0602-y
51. Xie YM, Steven GP (1993) A simple evolutionary procedure for structural optimization. Comput Struct 49:885–896. https://doi.org/10.1016/0045-7949(93)90035-C

# Chapter 5
# SSM-Based Optimization Framework

**Abstract** This chapter presents the Spectral Submanifold-based optimization framework developed to tailor the nonlinear dynamic response of mechanical systems. It describes the formulation, implementation, and main optimization objectives. A simple example involving a nonlinear oscillator chain illustrates its functionality.

## 5.1 Introduction

The optimization frameworks described in Chaps. 3 and 4 are typically applied in the context of linear dynamics. These approaches usually focus on maximizing or minimizing eigenfrequencies, or on shaping the linear frequency response of the system subjected to harmonic excitation. While effective in many engineering applications, such linear assumptions may fail to capture the actual behavior of the structure when nonlinearities become significant [14, 17].

A critical question thus arises: *what happens if the optimized design exhibits a nonlinear dynamic behavior*? This issue is especially relevant for structures at the micro-scale, such as MEMS devices, where geometric and material nonlinearities can strongly influence the system's response even under moderate excitation levels. In these regimes, linear approximations often fall short, leading to potentially suboptimal or even misleading design outcomes.

To address this gap, the ultimate objective of this research is to develop a computational framework for optimizing the nonlinear dynamic response of mechanical systems. The proposed approach would enable the systematic design of systems whose nonlinear behaviors can be tailored to meet specific performance targets.

Most of the available approaches in the literature (Table 5.1) rely on numerical methods, such as numerical continuation techniques [3] or the Harmonic Balance Method (HBM) [10], to approximate the nonlinear dynamic response of the system. While these methods are well-established, they are also computationally intensive, which limits their applicability to relatively small-scale models, particularly when used within optimization loops. Beyond their computational cost, these approaches

 
M. Pozzi, F. Braghin, *Structural Optimization for Nonlinear Dynamics*, PoliMI SpringerBriefs, https://doi.org/10.1007/978-3-032-22600-6_5

**Table 5.1** Overview of selected literature on structural optimization targeting nonlinear dynamic behavior. The table summarizes the type of nonlinearity considered, the specific dynamic target, the model reduction or analysis method employed, and the application domain

| | Nonlinearity | Reduction method |
|---|---|---|
| [7] | Geometric | Harmonic balance |
| [5] | Geometric | Normal forms |
| [6] | Geometric | Harmonic balance |
| [4] | Geometric | Harmonic balance |
| [13] | Material | Equivalent static loads |
| [2] | Geometric | Numerical continuation |
| [11] | Material | Numerical continuation |
| [17] | Geometric | Normal forms |
| [12] | Geometric | Stiffness evaluation procedure |
| | Target | Application |
| [7] | Resonance peak | Frame structure |
| [5] | Modal coupling coefficient | Beam resonator |
| [6] | Hardening coefficient | Frame structure |
| [4] | Backbone curve | 2 DOFs oscillator |
| [13] | Resonance peak | 3D structure |
| [2] | Nonlinear eigenvalue | Plane structure |
| [11] | Nonlinear eigenvalue | Plane structure |
| [17] | Modal coupling coefficient | MEMS gyroscope |
| [12] | Q-factor | NEMS resonator |

are highly sensitive to user-defined parameters, such as the number of harmonics in the HBM or the arc-length parameter in numerical continuation. As the system behavior may change significantly during the optimization process, these parameters often need to be adjusted, a task that is difficult to automate. Furthermore, inherent limitations of the numerical formulations, especially in the sensitivity analysis of HBM-based methods, can affect the convergence and robustness of the optimization [16].

## 5.2 Optimization Framework

To overcome these challenges, we rely on the Spectral Submanifold (SSM) theory to efficiently extract analytical expressions for the backbone and frequency response curves (Chap. 2). This approach significantly reduces dependence on user-defined parameters, with the expansion order being the only parameter that needs to be specified. The order of expansion can be automatically adjusted during the optimization to maintain the desired level of accuracy in the SSM-based reduction [15]. What is typically a drawback in numerical methods (parameter sensitivity) is thus transformed into a controlled mechanism that enhances both the robustness and

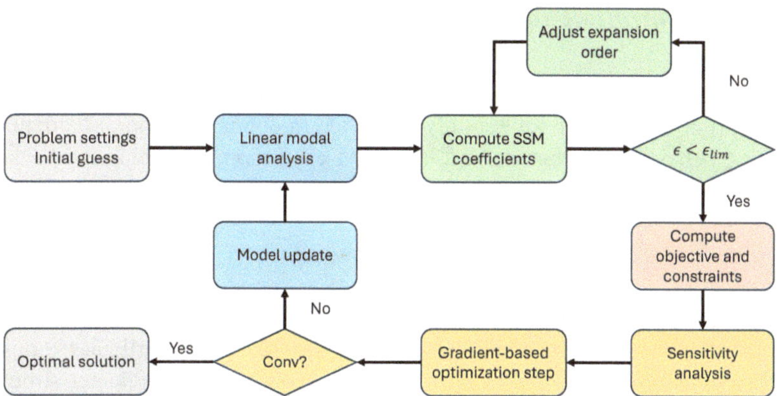


**Fig. 5.1** SSM-based optimization loop. Blue blocks correspond to the system model, while green blocks represent the SSM computation. The red block involves evaluating the objective function and constraints, which requires computing either the backbone curve or the frequency response curve, depending on the optimization formulation. Yellow blocks denote components related to the optimization process

convergence of the optimization process. At the first iteration, an accuracy threshold $\epsilon_{\text{max}}$ is selected based on a preliminary analysis of the system. This value represents an upper bound for the residual $\epsilon_{\text{inv}}$ of the invariance equation (2.19). At each iteration, the expansion order is automatically adjusted to ensure that $\epsilon_{\text{inv}} < \epsilon_{\text{max}}$. Additionally, a lower threshold $\epsilon_{\text{min}}$ can be specified to allow for a reduction in the expansion order when appropriate, thus reducing the computational cost while preserving the desired level of accuracy.

The SSM-based optimization loop (Fig. 5.1) begins with a linear modal analysis to compute the master eigenfrequency and corresponding mode shape. To ensure accurate mode tracking, the Modal Assurance Criterion (MAC) [9] can be employed. Once the target mode is identified, the SSM coefficients are computed up to the desired order of expansion.

Next, the accuracy inner loop is performed until the residual of the invariance equation falls below the specified threshold or the maximum expansion order is reached. Following this, the objective function and constraints are evaluated. This step involves computing either the backbone curve or the frequency response curve, depending on the specific optimization formulation.

The sensitivities of the objective function and constraints are then calculated and used to perform a gradient-based optimization step. Finally, convergence is checked. If the solution has not yet converged, the model is updated, and the loop is repeated. Otherwise, the optimal solution is obtained.

A MATLAB implementation of the overall optimization framework is implemented in `YetAnotherFEcode` [8].

### 5.2.1 Optimization Targets

The choice of optimization target is not unique and depends on the desired dynamic response. For example, if we are only interested in obtaining either a hardening or softening behavior, we can use an $\mathcal{O}(3)$ SSM expansion, for which the backbone curve is given as

$$\Omega = \mathrm{Im}\{\lambda\} + \mathrm{Im}\{\gamma_3\}\rho^2. \tag{5.1}$$

The third-order backbone coefficient $\mathrm{Im}\{\gamma_3\}$ governs the hardening/softening behavior of the response at small displacement amplitudes. In particular, a positive $\mathrm{Im}\{\gamma_3\}$ results in a hardening behavior, while a negative value indicates softening. Ideally, a linear behavior is obtained for $\mathrm{Im}\{\gamma_3\} = 0$, although achieving an exactly null value is practically unfeasible. Nonetheless, minimizing the absolute value of $\mathrm{Im}\{\gamma_3\}$ makes the response as linear as possible. This optimization target is particularly suitable in MEMS inertial sensors, where a nonlinear response is undesired and can negatively affect the device's performance. We refer to this class of problems as *hardening/softening optimization.*

Beyond the third-order coefficient, higher-order terms can also be included in the optimization to tailor the system's hardening or softening behavior at larger amplitudes. In general, the $M$-th order backbone coefficient $\gamma_M$ is obtained as

$$\gamma_M = R^1_{\boldsymbol{m}} = \bar{R}^2_{\boldsymbol{m}^s}, \tag{5.2}$$

where $\boldsymbol{m} = \{x, y\}$ and $\boldsymbol{m}^s = \{y, x\}$ with $x + y = M$ and $x - y = 1$.

While optimizing these coefficients can shape the desired nonlinear behavior, it does not provide direct control over the physical amplitude of the periodic response. As a result, the optimized system may exhibit the desired behavior only at amplitudes that are practically unfeasible for the specific application. Therefore, to tailor the frequency-amplitude relation in physical coordinates, we need to address the backbone curve directly. For example, we can optimize the system to have target frequencies at given amplitudes. We refer to this as *backbone optimization.*

Using the desired amplitude $x_{\mathrm{target}}$, the corresponding reduced amplitude $\rho$ is computed from

$$x(\rho) = \sqrt{\frac{1}{N_t}\sum_{k=1}^{N_t}(x^i_k(\rho))^2}, \tag{5.3}$$

by setting $x(\rho) = x_{\mathrm{target}}$.

The corresponding frequency is then evaluated using the computed value of $\rho$:

$$\Omega(\rho) = \mathrm{Im}\{\lambda\} + \sum_{m>1} \mathrm{Im}\{\gamma_m\}\rho^{m-1}. \tag{5.4}$$

This approach allows us to express the backbone frequency as a function of the physical amplitude. By selecting multiple frequency-amplitude targets, we can formulate an optimization problem to shape the backbone curve of mechanical systems. Additionally, the residual of the invariance equation (2.19) can be used to automatically adjust the expansion order to control the accuracy of the SSM-based reduction.

It is essential to carefully select the target amplitudes, as they may require unfeasible forcing levels, which typically depend on the specific application. For example, in MEMS devices, the external force is provided by electrostatic actuation [1], which depends on geometrical factors (e.g., the number and shape of comb fingers) and electrical parameters (e.g., voltage levels, capacitance). This information can be included in the optimization problem through the non-autonomous SSM coefficient $S_0$ defined in Sect. 2.3.2.

In this case, the optimization target is the peak of the frequency response at the selected forcing level $\varepsilon$. We refer to this as *frequency response peak optimization.* In this case, the reduced amplitude is computed from the peak condition

$$a^2(\rho) - \varepsilon^2 |S_0|^2 = 0, \tag{5.5}$$

where

$$a(\rho) = \sum_{\boldsymbol{m} \in \mathbb{N}^2} \mathrm{Re}\{\gamma_m\} \rho^m. \tag{5.6}$$

The computed value of $\rho$ is then used to evaluate the corresponding frequency and amplitude using Eqs. (5.3) and (5.4). Depending on the design requirements, the optimization objective may target the peak frequency, the peak amplitude, or both.

## 5.3 Optimization of a Nonlinear Oscillator Chain

To illustrate the functionality of the SSM-based optimization framework, we consider the nonlinear oscillator chain described in Sect. 2.7. The system is parametrized using the mass $m$ and the linear $k$, quadratic $k_2$, and cubic $k_3$ springs. The damping value $c = \alpha_R m + \beta_R k$ is computed using the Rayleigh damping with $\alpha_R = 0$ and $\beta_R = 0.1$. The initial values and bounds for the four design variables are shown in Table 5.2.

**Table 5.2** Initial values and bounds for the nonlinear oscillator chain

| | $m$ [kg] | $k$ [N/m] | $k_2$ [N/m$^2$] | $k_3$ [N/m$^2$] |
|---|---|---|---|---|
| Initial values | 3.00 | 30.0 | 3.00 | 3.00 |
| Lower bounds | 0.50 | 0.50 | 0.05 | 0.05 |
| Upper bounds | 20.0 | 50.0 | 75.0 | 75.0 |

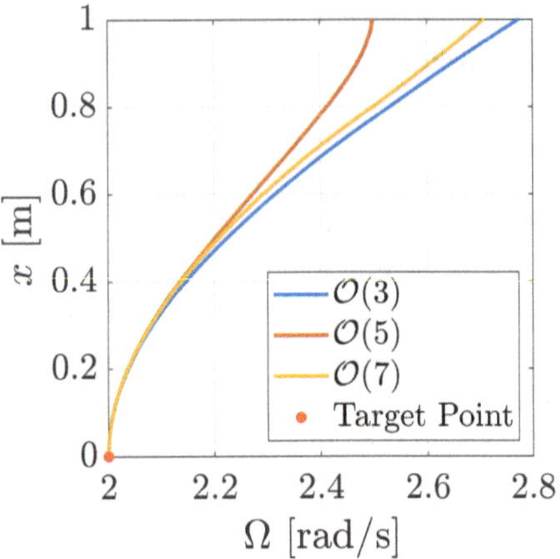


**Fig. 5.2** Backbone curve of the nonlinear oscillator chain optimized using problem (5.7)

For simplicity, a fixed third-order manifold expansion is used, and the sensitivities are computed following the adjoint method as described in Chap. 6.

We formulate the first optimization problem to impose the first eigenfrequency of the structure to 2 rad/s while maximizing the third-order backbone coefficient $\text{Im}\{\gamma_3\}$, which governs the hardening behavior of the system:

$$\begin{aligned} \max_{\boldsymbol{\mu}} \quad & \text{Im}\{\gamma_3\} \\ \text{s.t.} \quad & \omega - \omega_{\text{target}} = 0 \\ & \boldsymbol{\mu}_{\text{l}} \leq \boldsymbol{\mu} \leq \boldsymbol{\mu}_{\text{u}}, \end{aligned} \tag{5.7}$$

where $\text{Im}\{\gamma_3\}$ serves as the objective to maximize the system's hardening response. The backbone curve corresponding to the optimal design (Fig. 5.2) is evaluated at three expansion orders and exhibits a hardening behavior, with $\text{Im}\{\gamma_3\} = 0.97$.

To optimize for the frequency-amplitude relation of the response, the optimization problem (5.7) is modified to impose the target point (2.2 rad/s, 1 m) for the backbone curve:

$$\begin{aligned} \min_{\boldsymbol{\mu}} \quad & m \\ \text{s.t.} \quad & \omega - \omega_{\text{target}} = 0 \\ & \Omega(x_{\text{target}}) - \Omega_{\text{target}} = 0 \\ & \boldsymbol{\mu}_{\text{l}} \leq \boldsymbol{\mu} \leq \boldsymbol{\mu}_{\text{u}}, \end{aligned} \tag{5.8}$$

where the objective function is used to minimize the overall mass of the system.

The backbone curve of the optimized system is shown in Fig. 5.3, along with two frequency response curves obtained using different forcing levels. In particular, the force required to reach the target point is 4.28 N.

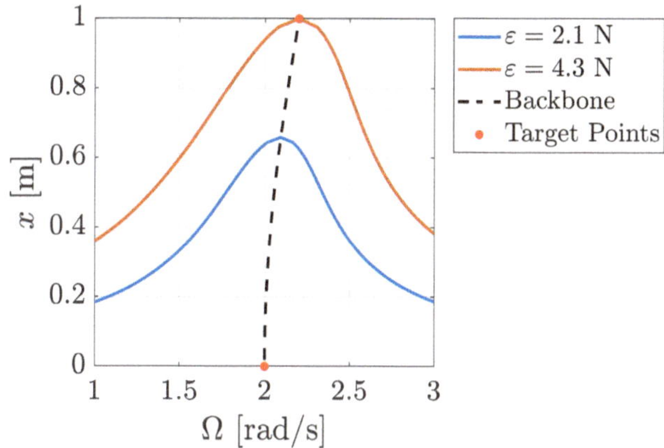


**Fig. 5.3** Frequency response of the nonlinear oscillator chain optimized using problem (5.8)

We now want to reach the same target point using half of the required force. To do so, we optimize for the frequency response peak corresponding to a forcing level $\varepsilon_{\text{target}} = 2.00\,\text{N}$. The optimization problem reads:

$$\begin{aligned} \min_{\boldsymbol{\mu}} \quad & m \\ \text{s.t.} \quad & \omega - \omega_{\text{target}} = 0 \\ & \Omega(\varepsilon_{\text{target}}) - \Omega_{\text{target}} = 0 \\ & x(\varepsilon_{\text{target}}) - x_{\text{target}} = 0 \\ & \boldsymbol{\mu}_{\text{l}} \leq \boldsymbol{\mu} \leq \boldsymbol{\mu}_{\text{u}}, \end{aligned} \tag{5.9}$$

where the objective function is used to minimize the overall mass of the system, and the second and third constraints are used to impose the frequency response peak corresponding to the prescribed force.

The frequency response of the optimal solution (Fig. 5.4) exhibits a significantly narrower resonance peak compared to that obtained from problem (5.8). Due to the constraint on the natural frequency, the optimal mass and linear stiffness values (Table 5.3) are identical in both cases. However, problem (5.9) yields different configurations for the nonlinear springs, resulting in a more compliant system that achieves the target response at a lower forcing amplitude.

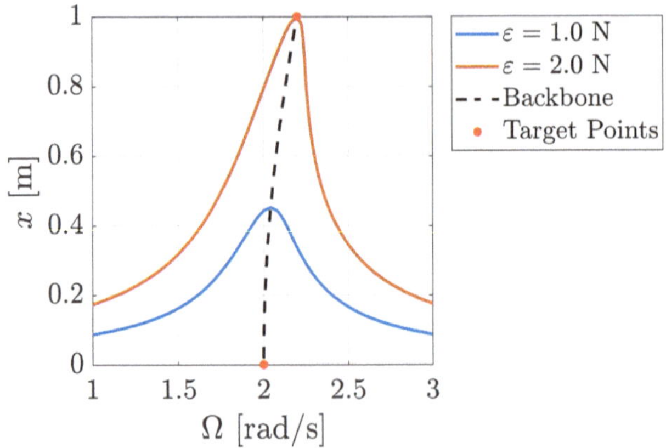


**Fig. 5.4** Frequency response of the nonlinear oscillator chain optimized using problem (5.9)

**Table 5.3** Optimal solutions for the optimization problems (5.7)–(5.9)

| | $m$ [kg] | $k$ [N/m] | $k_2$ [N/m$^2$] | $k_3$ [N/m$^2$] |
|---|---|---|---|---|
| Problem (5.7) | 0.50 | 13.3 | 0.20 | 75.0 |
| Problem (5.8) | 0.50 | 13.3 | 15.5 | 18.1 |
| Problem (5.9) | 0.50 | 13.3 | 1.04 | 16.6 |

## References


1. Acar C, Shkel A (2009) MEMS vibratory gyroscopes. Springer, Berlin. https://doi.org/10.1007/978-0-387-09536-3
2. Dalklint A, Wallin M, Tortorelli DA (2020) Eigenfrequency constrained topology optimization of finite strain hyperelastic structures. Struct Multidiscip Optim 61:2577–2594. https://doi.org/10.1007/s00158-020-02557-9
3. Dankowicz H, Schilder F (2013) Recipes for continuation. Society for Industrial and Applied Mathematics, Philadelphia. https://doi.org/10.1137/1.9781611972573
4. Detroux T, Noël JP, Kerschen G (2021) Tailoring the resonances of nonlinear mechanical systems. Nonlinear Dyn 103:3611–3624. https://doi.org/10.1007/s11071-020-06002-w
5. Dou S, Jensen JS (2015) Optimization of nonlinear structural resonance using the incremental harmonic balance method. J Sound Vib 334:239–254. https://doi.org/10.1016/j.jsv.2014.08.023
6. Dou S, Jensen JS (2016) Optimization of hardening/softening behavior of plane frame structures using nonlinear normal modes. Comput Struct 164:63–74. https://doi.org/10.1016/j.compstruc.2015.11.001
7. Dou S, Strachan BS, Shaw SW, Jensen JS (2015) Structural optimization for nonlinear dynamic response. Philos Trans R Soc A Math Phys Eng Sci 373:20140408. https://doi.org/10.1098/rsta.2014.0408
8. Jain S, Marconi J, Pozzi M, Tiso P (2026) YetAnotherFEcode (v1.5.0). Zenodo. https://doi.org/10.5281/zenodo.15257966
9. Kim TS, Kim YY (2002) Mac-based mode-tracking in structural topology optimization. Comput Struct 74:375–383. https://doi.org/10.1016/S0045-7949(99)00056-5
10. Krack M, Gross J (2019) Harmonic balance for nonlinear vibration problems. Springer, Berlin (2019). https://doi.org/10.1007/978-3-030-14023-6

11. Li Q, Wu Q, Dou S, Wang J, Liu S, Chen W (2022) Nonlinear eigenvalue topology optimization for structures with frequency-dependent material properties. Mech Syst Signal Process 170:108835. https://doi.org/10.1016/j.ymssp.2022.108835
12. Li Z, Alijani F, Sarafraz A, Xu M, Norte RA, Aragón AM, Steeneken PG (2025) Finite element-based nonlinear dynamic optimization of nanomechanical resonators. Microsyst Nanoeng 11:16. https://doi.org/10.1038/s41378-024-00854-7
13. Lu S, Zhang Z, Guo H, Park GJ, Zuo W (2021) Nonlinear dynamic topology optimization with explicit and smooth geometric outline via moving morphable components method. Struct Multidiscip Optim 64:2465–2487. https://doi.org/10.1007/s00158-021-03000-3
14. Marconi J, Bonaccorsi G, Giannini D, Falorni L, Braghin F (2021) Exploiting nonlinearities for frequency-matched mems gyroscopes tuning. In: 2021 IEEE international symposium on inertial sensors and systems (INERTIAL). IEEE, Piscataway, pp 1–4. https://doi.org/10.1109/INERTIAL51137.2021.9430478
15. Pozzi M, Marconi J, Jain S, Braghin F (2024) Backbone curve tailoring via lyapunov subcenter manifold optimization. Nonlinear Dyn. 112:15719–15739. https://doi.org/10.1007/s11071-024-09881-5
16. Saccani A, Marconi J, Tiso P (2023) Sensitivity analysis of nonlinear frequency response of defected structures. Nonlinear Dyn. 111:4027–4051. https://doi.org/10.1007/s11071-022-08064-4
17. Schiwietz D, Hörsting M, Weig EM, Wenzel M, Degenfeld-Schonburg P (2025) Shape optimization of geometrically nonlinear modal coupling coefficients: an application to mems gyroscopes. Sci Rep 15:10957. https://doi.org/10.1038/s41598-025-95412-0

# Chapter 6
# Sensitivity Analysis

**Abstract** This chapter derives sensitivity expressions for frequency response and backbone curves computed using SSM theory. Two approaches are discussed—the direct differentiation method and the adjoint method. To demonstrate their practical relevance beyond optimization, a representative example is provided to quantify how parameter perturbations affect system response.

## 6.1 Introduction

In structural optimization, the sensitivities of the objective function and constraints are required for solving optimization problems using gradient-based methods, including interior-point and dual-simplex algorithms [1]. Two primary approaches are commonly employed to compute these sensitivities: the direct differentiation method and the adjoint method [2]. In this chapter, we apply both methods to derive sensitivity expressions for the optimization targets defined in Chap. 5, i.e. the $M$-th order backbone coefficient $\mathrm{Im}\{\gamma_M\}$, the response frequency $\Omega$, and the response amplitude $x$.

To compute the sensitivities, the expression for $\mathrm{Im}\{\gamma_M\}$ in Eq. (5.2) can be rewritten as

$$\mathrm{Im}\{\gamma_M\} = \frac{1}{2}\mathrm{i}\left(\gamma_M^2 - \gamma_M^1\right). \tag{6.1}$$

The same is done for the frequency $\Omega$ in Eq. (5.4), which is rewritten as

$$\Omega = \frac{1}{2}\mathrm{i}\left(\lambda_2 - \lambda_1\right) + \frac{1}{2}\mathrm{i}\sum_{m>1}\left(\gamma_m^2 - \gamma_m^1\right)\rho^{m-1}. \tag{6.2}$$



M. Pozzi, F. Braghin, *Structural Optimization for Nonlinear Dynamics*,
PoliMI SpringerBriefs, https://doi.org/10.1007/978-3-032-22600-6_6

## 6.2 Direct Differentiation Sensitivity

The direct differentiation approach for computing sensitivities involves applying the chain rule to differentiate all quantities involved in computing the function of interest. For example, the sensitivity of the $M$-th order backbone coefficient $\mathrm{Im}\{\gamma_M\}$ is:

$$\frac{\mathrm{d}\,\mathrm{Im}\{\gamma_M\}}{\mathrm{d}\mu} = \frac{1}{2}\mathrm{i}\left(\frac{\mathrm{d}\gamma_M^2}{\mathrm{d}\mu} - \frac{\mathrm{d}\gamma_M^1}{\mathrm{d}\mu}\right). \tag{6.3}$$

Similarly, the sensitivity of the response frequency of the backbone curve is obtained by differentiating Eq. (6.2):

$$\begin{aligned}\frac{\mathrm{d}\Omega}{\mathrm{d}\mu} &= \frac{1}{2}\mathrm{i}\left(\frac{\mathrm{d}\lambda_2}{\mathrm{d}\mu} - \frac{\mathrm{d}\lambda_1}{\mathrm{d}\mu}\right) \\ &+ \frac{1}{2}\mathrm{i}\sum_{m>1}\left[\left(\frac{\mathrm{d}\gamma_m^2}{\mathrm{d}\mu} - \frac{\mathrm{d}\gamma_m^1}{\mathrm{d}\mu}\right)\rho^{m-1}\right. \\ &\left. + \frac{\mathrm{d}\rho}{\mathrm{d}\mu}\left(\gamma_m^2 - \gamma_m^1\right)(m-1)\rho^{m-2}\right],\end{aligned} \tag{6.4}$$

where this expression depends on the derivative of the reduced amplitude $\rho$, the eigenvalues $\lambda_j$, and all the non-zero reduced dynamic coefficients $R_{\boldsymbol{m}}^j$ used to compute the backbone coefficients $\gamma_m^j$.

The sensitivity of the response amplitude in Eq. (5.3) is

$$\frac{\mathrm{d}x}{\mathrm{d}\mu} = \frac{1}{xN_t}\sum_{k=1}^{N_t} x_k^i \frac{\mathrm{d}x_k^i}{\mathrm{d}\mu}, \tag{6.5}$$

where $x_k^i(\rho) = x^i(\rho, \theta_k)$ and

$$\frac{\mathrm{d}x_k^i}{\mathrm{d}\mu} = \sum_{\boldsymbol{m}\in\mathbb{N}^2} \frac{\mathrm{d}w_{\boldsymbol{m}}^i}{\mathrm{d}\mu}\boldsymbol{p}_k^{\boldsymbol{m}} + \frac{\mathrm{d}\rho}{\mathrm{d}\mu}\sum_{\boldsymbol{m}\in\mathbb{N}^2} w_{\boldsymbol{m}}^i m\rho^{m-1}\tilde{\boldsymbol{p}}_k^{\boldsymbol{m}}, \tag{6.6}$$

with

$$\boldsymbol{p}_k = \rho\tilde{\boldsymbol{p}}_k = \rho\left[e^{\mathrm{i}\theta_k}, e^{-\mathrm{i}\theta_k}\right]^T. \tag{6.7}$$

There are two possibilities for computing the derivative of $\rho$, depending on the optimization problem formulation. If we want to impose the response frequency at a given displacement amplitude, then the derivative of $\rho$ is obtained from Eq. (6.5) and (6.6) by setting $\frac{\mathrm{d}x}{\mathrm{d}\mu} = 0$:

$$\frac{\mathrm{d}\rho}{\mathrm{d}\mu} = -\frac{\sum_{\boldsymbol{m}\in\mathbb{N}^2} \frac{\mathrm{d}w^i_{\boldsymbol{m}}}{\mathrm{d}\mu} \boldsymbol{p}_k^{\boldsymbol{m}}}{\sum_{\boldsymbol{m}\in\mathbb{N}^2} w^i_{\boldsymbol{m}} m\rho^{m-1} \tilde{\boldsymbol{p}}_k^{\boldsymbol{m}}}. \tag{6.8}$$

In this case, the non-autonomous contribution of the SSM is not required, and the backbone curve is computed from the autonomous SSM, thus neglecting the effect of the external force. Otherwise, if we want to optimize the frequency, the amplitude, or both for a given forcing level $\varepsilon$, the non-autonomous SSM contribution is required and the derivative of $\rho$ is computed by differentiating Eq. (5.5), leading to

$$\frac{\mathrm{d}\rho}{\mathrm{d}\mu} = \frac{\varepsilon^2 \operatorname{Re}\left\{\bar{S}_0 \frac{\mathrm{d}S_0}{\mathrm{d}\mu}\right\}/a(\rho) - \sum_{m\in\mathbb{N}} \left(\frac{\mathrm{d}\gamma^2_m}{\mathrm{d}\mu} + \frac{\mathrm{d}\gamma^1_m}{\mathrm{d}\mu}\right)\rho^m}{\sum_{m\in\mathbb{N}} \left(\gamma^2_m + \gamma^1_m\right) m\rho^{m-1}}. \tag{6.9}$$

To enhance readability, all remaining total derivatives appearing in Eqs. (6.3)–(6.9) are collected at the end of the chapter (Sect. 6.5).

### 6.2.1 *Finite Differences Validation*

To ensure the accuracy of the sensitivity computation, the sensitivities obtained via direct differentiation are validated against those obtained via finite differences. This comparison is performed for multiple SSM expansion orders, from $\mathcal{O}(3)$ to $\mathcal{O}(7)$, and for a range of perturbation step sizes $\delta\mu$.

For each design variable $\mu$, the finite difference approximation is computed using the forward difference formula:

$$\frac{\mathrm{d}J}{\mathrm{d}\mu} \approx \frac{J(\mu + \delta\mu) - J(\mu)}{\delta\mu}, \tag{6.10}$$

where $J$ is the quantity of interest and $\delta\mu$ is the perturbation step.

Figure 6.1 illustrates the sensitivity error as a function of the step size for the four different functions of interest described in Sect. 6.2, i.e. the $\mathcal{O}(3)$ backbone coefficient $\operatorname{Im}\{\gamma_3\}$, the backbone frequency, and the frequency and amplitude of the response peak. The small relative error confirms the correctness of the analytical derivation and implementation.

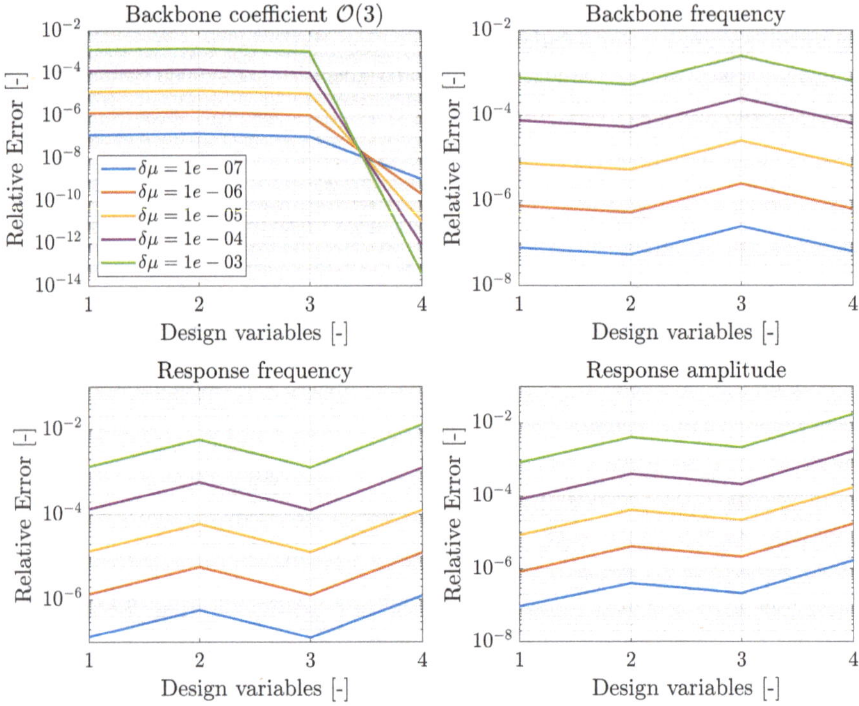


**Fig. 6.1** Sensitivity validation against finite differences. The plot shows the relative error between direct differentiation sensitivities and finite difference approximations for different perturbation step sizes

## 6.3 Adjoint Sensitivity

To efficiently handle optimization problems with a large number of design variables, we compute the sensitivities using the adjoint method [2], in which the complex derivatives are managed through *Wirtinger calculus* [2, 4].

In this section, we derive the adjoint sensitivities for the $M$-th order reduced dynamic coefficients $\text{Im}\{\gamma_M\}$, the response frequency $\Omega$, and the response amplitude $x$. To enhance readability, we describe here only the Lagrangian, the adjoint equations, and the sensitivity expression, while all the other partial derivatives appearing in Eqs. (6.11)–(6.19) are collected at the end of the chapter (Sect. 6.6).

### *6.3.1 Backbone Coefficient Adjoint Sensitivity*

The first step in computing the adjoint sensitivity is the identification of the state variables and functions, which are used, together with the adjoint variables, to assemble the Lagrangian. For example, the state variables involved in the computation of $\mathrm{Im}\{\gamma_M\}$ are the master eigenfrequency $\omega$, the corresponding mode shape $\boldsymbol{\varphi}$, and the manifold coefficients $\boldsymbol{w_m}$ at orders lower than $M$. The Lagrangian is then assembled as

$$\mathcal{L} = \mathrm{Im}\{\gamma_M\} + \boldsymbol{\lambda}_\varphi^T \left(\mathbf{K} - \omega^2\mathbf{M}\right) \boldsymbol{\varphi} + \lambda_\omega \left(\boldsymbol{\varphi}^T \mathbf{M}\boldsymbol{\varphi} - 1\right) + \sum_{\substack{\boldsymbol{m}\in\mathbb{N}^2 \\ 1<|\boldsymbol{m}|<M}} \boldsymbol{\lambda}_{\boldsymbol{m}}^T \left(\mathbf{L}_{\boldsymbol{m}} \boldsymbol{w}_{\boldsymbol{m}} - \boldsymbol{h}_{\boldsymbol{m}}\right), \tag{6.11}$$

where $\boldsymbol{\lambda_m} \in \mathbb{C}^n$, $\boldsymbol{\lambda}_\varphi \in \mathbb{R}^n$, and $\lambda_\omega \in \mathbb{R}$ are the adjoint variables associated with the state equations. Since the Lagrangian must be real-valued [2, 4], only the real part of the cohomological equations (2.21) should be used. However, the *symmetric multi-index* property (Sect. 2.3.1) ensures that these equations naturally occur in complex-conjugate pairs, making their sum inherently real. As a consequence, the adjoint variables associated with symmetric multi-indices are also complex conjugates, which guarantees the real-valued nature of the Lagrangian [3].

The adjoint equations are obtained by setting to zero the partial derivatives of $\mathcal{L}$ with respect to the state variables:

$$\frac{\partial \mathcal{L}}{\partial \boldsymbol{w}_{\boldsymbol{m}}} = \frac{\partial\, \mathrm{Im}\{\gamma_M\}}{\partial \boldsymbol{w}_{\boldsymbol{m}}} + \boldsymbol{\lambda}_{\boldsymbol{m}}^T \mathbf{L}_{\boldsymbol{m}} - \sum_{\substack{\boldsymbol{q}\in\mathbb{N}^2 \\ |\boldsymbol{q}|>|\boldsymbol{m}|}} \boldsymbol{\lambda}_{\boldsymbol{q}}^T \frac{\partial \boldsymbol{h}_{\boldsymbol{q}}}{\partial \boldsymbol{w}_{\boldsymbol{m}}} = \mathbf{0}^T, \quad \forall \boldsymbol{m} : 1 < |\boldsymbol{m}| < M \tag{6.12a}$$

$$\frac{\partial \mathcal{L}}{\partial \boldsymbol{\varphi}} = \frac{\partial\, \mathrm{Im}\{\gamma_M\}}{\partial \boldsymbol{\varphi}} + \boldsymbol{\lambda}_\varphi^T \left(\mathbf{K} - \omega^2\mathbf{M}\right) + 2\lambda_\omega \boldsymbol{\varphi}^T \mathbf{M} - \sum_{\substack{\boldsymbol{m}\in\mathbb{N}^2 \\ 1<|\boldsymbol{m}|<M}} \boldsymbol{\lambda}_{\boldsymbol{m}}^T \frac{\partial \boldsymbol{h}_{\boldsymbol{m}}}{\partial \boldsymbol{\varphi}} = \mathbf{0}^T \tag{6.12b}$$

$$\frac{\partial \mathcal{L}}{\partial \omega} = \frac{\partial\, \mathrm{Im}\{\gamma_M\}}{\partial \omega} - 2\omega \boldsymbol{\lambda}_\varphi^T \mathbf{M}\boldsymbol{\varphi} + \sum_{\substack{\boldsymbol{m}\in\mathbb{N}^2 \\ 1<|\boldsymbol{m}|<M}} \boldsymbol{\lambda}_{\boldsymbol{m}}^T \left(\frac{\partial \mathbf{L}_{\boldsymbol{m}}}{\partial \omega} \boldsymbol{w}_{\boldsymbol{m}} - \frac{\partial \boldsymbol{h}_{\boldsymbol{m}}}{\partial \omega}\right) = 0, \tag{6.12c}$$

where the first equation is solved backward for each multi-index $\boldsymbol{m}$ at every order, starting from the highest one, and the last two are solved together in the same linear system. Note that the matrices $\mathbf{L_m}$ depend only on $\omega$ and $\mu$, while the vectors $\boldsymbol{h_m}$ also depend on $\boldsymbol{\varphi}$ and on the manifold coefficients $\boldsymbol{w_q}$ at lower orders (Sect. 2.3.1).

The sensitivity of $\mathrm{Im}\{\gamma_M\}$ is then the partial derivative of $\mathcal{L}$ with respect to $\mu$:

$$\begin{aligned}\frac{\partial \mathcal{L}}{\partial \mu} = \frac{\partial\, \mathrm{Im}\{\gamma_M\}}{\partial \mu} + \boldsymbol{\lambda}_{\varphi}^{T} \left( \frac{\partial \mathbf{K}}{\partial \mu} - \omega^2 \frac{\partial \mathbf{M}}{\partial \mu} \right) \boldsymbol{\varphi} + \lambda_{\omega} \boldsymbol{\varphi}^T \frac{\partial \mathbf{M}}{\partial \mu} \boldsymbol{\varphi} \\ + \sum_{\substack{\boldsymbol{m} \in \mathbb{N}^2 \\ 1<|\boldsymbol{m}|<M}} \boldsymbol{\lambda}_{\boldsymbol{m}}^{T} \left( \frac{\partial \mathbf{L}_{\boldsymbol{m}}}{\partial \mu} \boldsymbol{w}_{\boldsymbol{m}} - \frac{\partial \boldsymbol{h}_{\boldsymbol{m}}}{\partial \mu} \right).\end{aligned} \tag{6.13}$$

### 6.3.2 Backbone Frequency Adjoint Sensitivity

We apply the adjoint method to compute the sensitivity of the backbone frequency $\Omega$ for a given displacement amplitude $x(\rho) = x_{\text{target}}$. Similarly to Eq. (6.11), the Lagrangian is assembled as

$$\begin{aligned}\mathcal{L} = \Omega + \boldsymbol{\lambda}_{\varphi}^{T} \left( \mathbf{K} - \omega^2 \mathbf{M} \right) \boldsymbol{\varphi} + \lambda_{\omega} \left( \boldsymbol{\varphi}^T \mathbf{M} \boldsymbol{\varphi} - 1 \right) \\ + \sum_{\substack{\boldsymbol{m} \in \mathbb{N}^2 \\ |\boldsymbol{m}|>1}} \boldsymbol{\lambda}_{\boldsymbol{m}}^{T} \left( \mathbf{L}_{\boldsymbol{m}} \boldsymbol{w}_{\boldsymbol{m}} - \boldsymbol{h}_{\boldsymbol{m}} \right) + \lambda_{\rho} \left( x - x_{\text{target}} \right),\end{aligned} \tag{6.14}$$

where $\lambda_\rho \in \mathbb{R}$, $\boldsymbol{\lambda}_{\boldsymbol{m}} \in \mathbb{C}^n$, $\boldsymbol{\lambda}_{\varphi} \in \mathbb{R}^n$, and $\lambda_\omega \in \mathbb{R}$ are the adjoint variables associated with the state variables $\rho$, $\boldsymbol{w}_{\boldsymbol{m}}$, $\boldsymbol{\varphi}$, and $\omega$.

The corresponding adjoint equations are:

$$\frac{\partial \mathcal{L}}{\partial \rho} = \frac{\partial \Omega}{\partial \rho} + \lambda_\rho \frac{\partial x}{\partial \rho} = 0 \tag{6.15a}$$

$$\begin{aligned}\frac{\partial \mathcal{L}}{\partial \boldsymbol{w}_{\boldsymbol{m}}} = \frac{\partial \Omega}{\partial \boldsymbol{w}_{\boldsymbol{m}}} + \boldsymbol{\lambda}_{\boldsymbol{m}}^{T} \mathbf{L}_{\boldsymbol{m}} - \sum_{\substack{\boldsymbol{q} \in \mathbb{N}^2 \\ |\boldsymbol{q}|>|\boldsymbol{m}|}} \boldsymbol{\lambda}_{\boldsymbol{q}}^{T} \frac{\partial \boldsymbol{h}_{\boldsymbol{q}}}{\partial \boldsymbol{w}_{\boldsymbol{m}}} \\ + \lambda_\rho \frac{\partial x}{\partial \boldsymbol{w}_{\boldsymbol{m}}} = \mathbf{0}^T, \quad \forall \boldsymbol{m} : |\boldsymbol{m}| > 1\end{aligned} \tag{6.15b}$$

$$\begin{aligned}\frac{\partial \mathcal{L}}{\partial \boldsymbol{\varphi}} = \frac{\partial \Omega}{\partial \boldsymbol{\varphi}} + \boldsymbol{\lambda}_{\varphi}^{T} \left( \mathbf{K} - \omega^2 \mathbf{M} \right) + 2 \lambda_\omega \boldsymbol{\varphi}^T \mathbf{M} \\ - \sum_{\substack{\boldsymbol{m} \in \mathbb{N}^2 \\ |\boldsymbol{m}|>1}} \boldsymbol{\lambda}_{\boldsymbol{m}}^{T} \frac{\partial \boldsymbol{h}_{\boldsymbol{m}}}{\partial \boldsymbol{\varphi}} + \lambda_\rho \frac{\partial x}{\partial \boldsymbol{\varphi}} = \mathbf{0}^T\end{aligned} \tag{6.15c}$$

$$\frac{\partial \mathcal{L}}{\partial \omega} = \frac{\partial \Omega}{\partial \omega} - 2\omega \boldsymbol{\lambda}_\varphi^T \mathbf{M} \boldsymbol{\varphi} + \sum_{\substack{\boldsymbol{m} \in \mathbb{N}^2 \\ |\boldsymbol{m}|>1}} \boldsymbol{\lambda}_{\boldsymbol{m}}^T \left( \frac{\partial \mathbf{L}_{\boldsymbol{m}}}{\partial \omega} \boldsymbol{w}_{\boldsymbol{m}} - \frac{\partial \boldsymbol{h}_{\boldsymbol{m}}}{\partial \omega} \right) + \lambda_\rho \frac{\partial x}{\partial \boldsymbol{\varphi}} = 0, \tag{6.15d}$$

where the second equation is solved backward for each multi-index $\boldsymbol{m}$ at every order, starting from the highest one, and the last two are solved together in the same linear system. The sensitivity of $\Omega$ is then the partial derivative of $\mathcal{L}$ with respect to $\mu$:

$$\frac{\partial \mathcal{L}}{\partial \mu} = \frac{\partial \Omega}{\partial \mu} + \boldsymbol{\lambda}_\varphi^T \left( \frac{\partial \mathbf{K}}{\partial \mu} - \omega^2 \frac{\partial \mathbf{M}}{\partial \mu} \right) \boldsymbol{\varphi} + \lambda_\omega \boldsymbol{\varphi}^T \frac{\partial \mathbf{M}}{\partial \mu} \boldsymbol{\varphi} + \sum_{\substack{\boldsymbol{m} \in \mathbb{N}^2 \\ |\boldsymbol{m}|>1}} \boldsymbol{\lambda}_{\boldsymbol{m}}^T \left( \frac{\partial \mathbf{L}_{\boldsymbol{m}}}{\partial \mu} \boldsymbol{w}_{\boldsymbol{m}} - \frac{\partial \boldsymbol{h}_{\boldsymbol{m}}}{\partial \mu} \right) + \lambda_\rho \frac{\partial x}{\partial \mu}. \tag{6.16}$$

### 6.3.3 Frequency Response Adjoint Sensitivity

In Sect. 6.3.2, the reduced amplitude $\rho$ is computed by imposing a reference displacement amplitude $x(\rho) = x_{\text{target}}$. In this section, we compute $\rho$ for a given forcing level by solving Eq. (5.5), which becomes the state equation associated with $\rho$. In this case, both $\Omega$ and $x$ share the same state variables and functions, thus leading to the same Lagrangian formulation:

$$\mathcal{L} = J + \boldsymbol{\lambda}_\varphi^T \left( \mathbf{K} - \omega^2 \mathbf{M} \right) \boldsymbol{\varphi} + \lambda_\omega \left( \boldsymbol{\varphi}^T \mathbf{M} \boldsymbol{\varphi} - 1 \right) + \sum_{\substack{\boldsymbol{m} \in \mathbb{N}^2 \\ |\boldsymbol{m}|>1}} \boldsymbol{\lambda}_{\boldsymbol{m}}^T \left( \mathbf{L}_{\boldsymbol{m}} \boldsymbol{w}_{\boldsymbol{m}} - \boldsymbol{h}_{\boldsymbol{m}} \right) + \lambda_\rho \left( a^2(\rho) - \varepsilon^2 |S_0|^2 \right), \tag{6.17}$$

where $J$ is a placeholder use to indicate either $\Omega$ or $x$.

The corresponding adjoint equations are

$$\frac{\partial \mathcal{L}}{\partial \rho} = \frac{\partial J}{\partial \rho} + \lambda_\rho 2a \frac{\partial a}{\partial \rho} = 0 \tag{6.18a}$$

$$\frac{\partial \mathcal{L}}{\partial \boldsymbol{w}_m} = \frac{\partial J}{\partial \boldsymbol{w}_m} + \boldsymbol{\lambda}_{\boldsymbol{m}}^T \mathbf{L}_{\boldsymbol{m}} - \sum_{\substack{\boldsymbol{q} \in \mathbb{N}^2 \\ |\boldsymbol{q}|>|\boldsymbol{m}|}} \boldsymbol{\lambda}_{\boldsymbol{q}}^T \frac{\partial \boldsymbol{h}_{\boldsymbol{q}}}{\partial \boldsymbol{w}_{\boldsymbol{m}}} + \lambda_\rho 2a \frac{\partial a}{\partial \boldsymbol{w}_{\boldsymbol{m}}} = \mathbf{0}^T, \quad \forall \boldsymbol{m} : |\boldsymbol{m}| > 1 \tag{6.18b}$$

$$\frac{\partial \mathcal{L}}{\partial \boldsymbol{\varphi}} = \frac{\partial J}{\partial \boldsymbol{\varphi}} + \boldsymbol{\lambda}_{\varphi}^{T}\left(\mathbf{K} - \omega^{2}\mathbf{M}\right) + 2\lambda_{\omega}\boldsymbol{\varphi}^{T}\mathbf{M}$$
$$- \sum_{\substack{\boldsymbol{m}\in\mathbb{N}^{2}\\ |\boldsymbol{m}|>1}} \boldsymbol{\lambda}_{\boldsymbol{m}}^{T}\frac{\partial \boldsymbol{h}_{\boldsymbol{m}}}{\partial \boldsymbol{\varphi}} + \lambda_{\rho}\frac{\partial x}{\partial \boldsymbol{\varphi}} \tag{6.18c}$$
$$+ \lambda_{\rho}\left(2a\frac{\partial a}{\partial \boldsymbol{\varphi}} - \varepsilon^{2}2\,\mathrm{Re}\left\{\frac{\partial S_{0}}{\partial \boldsymbol{\varphi}}\bar{S}_{0}\right\}\right) = \mathbf{0}^{T}$$

$$\frac{\partial \mathcal{L}}{\partial \omega} = \frac{\partial J}{\partial \omega} - 2\omega\boldsymbol{\lambda}_{\varphi}^{T}\mathbf{M}\boldsymbol{\varphi}$$
$$+ \sum_{\substack{\boldsymbol{m}\in\mathbb{N}^{2}\\ |\boldsymbol{m}|>1}} \boldsymbol{\lambda}_{\boldsymbol{m}}^{T}\left(\frac{\partial \mathbf{L}_{\boldsymbol{m}}}{\partial \omega}\boldsymbol{w}_{\boldsymbol{m}} - \frac{\partial \boldsymbol{h}_{\boldsymbol{m}}}{\partial \omega}\right) \tag{6.18d}$$
$$+ \lambda_{\rho}\left(2a\frac{\partial a}{\partial \omega} - \varepsilon^{2}2\,\mathrm{Re}\left\{\frac{\partial S_{0}}{\partial \omega}\bar{S}_{0}\right\}\right) = 0,$$

where the second equation is solved backward for each multi-index $\boldsymbol{m}$ at every order, starting from the highest one, and the last two are solved together in the same linear system. The sensitivity of $J$ is then the partial derivative of $\mathcal{L}$ with respect to $\mu$:

$$\frac{\partial \mathcal{L}}{\partial \mu} = \frac{\partial J}{\partial \mu} + \boldsymbol{\lambda}_{\varphi}^{T}\left(\frac{\partial \mathbf{K}}{\partial \mu} - \omega^{2}\frac{\partial \mathbf{M}}{\partial \mu}\right)\boldsymbol{\varphi} + \lambda_{\omega}\boldsymbol{\varphi}^{T}\frac{\partial \mathbf{M}}{\partial \mu}\boldsymbol{\varphi}$$
$$+ \sum_{\substack{\boldsymbol{m}\in\mathbb{N}^{2}\\ |\boldsymbol{m}|>1}} \boldsymbol{\lambda}_{\boldsymbol{m}}^{T}\left(\frac{\partial \mathbf{L}_{\boldsymbol{m}}}{\partial \mu}\boldsymbol{w}_{\boldsymbol{m}} - \frac{\partial \boldsymbol{h}_{\boldsymbol{m}}}{\partial \mu}\right) + \lambda_{\rho}\frac{\partial x}{\partial \mu} \tag{6.19}$$
$$+ \lambda_{\rho}\left(2a\frac{\partial a}{\partial \mu} - \varepsilon^{2}2\,\mathrm{Re}\left\{\frac{\partial S_{0}}{\partial \mu}\bar{S}_{0}\right\}\right).$$

Equation (6.19) can be used to obtain the sensitivities of $\Omega$ and $x$ by substituting $J = \Omega$ or $J = x$, respectively. The partial derivatives of both cases are reported in Sect. 6.6.

### *6.3.4 Computational Time*

The adjoint-based sensitivities are validated against those obtained via direct differentiation, which have been benchmarked against finite differences (Sect. 6.2.1). While both methods yield identical sensitivity values, their computational performance differs significantly.

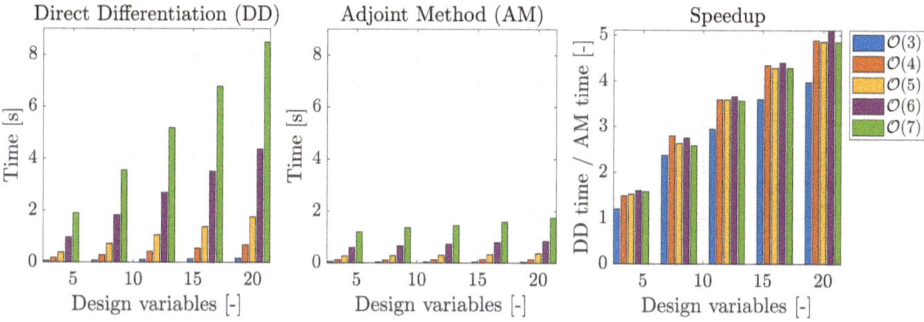


**Fig. 6.2** Computational times for the sensitivity analysis of a system with 501 degrees of freedom. Left: direct differentiation (DD). Center: adjoint method (AM). Right: speedup (ratio of DD time to AM time). The methods are compared across different SSM expansion orders and numbers of design variables

Figure 6.2 presents the computational times required for the sensitivity analysis of a system with 501 degrees of freedom. The performance of the adjoint formulation is compared against the direct differentiation approach for different numbers of design variables and SSM expansion orders. All simulations were performed on a Windows workstation equipped with an Intel Core i9-13900KS CPU @ 3.20 GHz and 128 GB of RAM @ 4200 MT/s.

As anticipated, the computational time of direct differentiation scales linearly with the number of design variables, since one linear system must be solved per parameter. In contrast, the adjoint method requires solving one adjoint system per state variable, regardless the number of design variables.[1]

## 6.4 Example of Sensitivity Analysis

Beyond optimization, the sensitivities are also employed to analyze how perturbations in the system parameters affect the system's response, a process commonly referred to as *sensitivity analysis*.

As an illustrative example, consider a system of two coupled oscillators with quadratic and cubic nonlinear springs. The governing equations of motion are given by:

[1] The adjoint sensitivity expression in Eqs. (6.13), (6.16), and (6.19) must be evaluated for each design variable, introducing a minor linear overhead. However, this is negligible compared to the linear scaling of direct differentiation.

$$
\begin{aligned}
m\ddot{x}_1 + c\dot{x}_1 + k(2x_1 - x_2) + k_2 x_1^2 + k_3 x_1^3 \\
+ k_2(x_1 - x_2)^2 + k_3(x_1 - x_2)^3 = 0, \\
m\ddot{x}_2 + c\dot{x}_2 + k(x_2 - x_1) \\
+ k_2(x_2 - x_1)^2 + k_3(x_2 - x_1)^3 = 0,
\end{aligned}
\tag{6.20}
$$

where the system parameters are set to $m = 1\,\text{kg}$, $k = 1\,\text{N/m}$, $k_2 = 0.5\,\text{N/m}^2$, and $k_3 = 0.2\,\text{N/m}^3$. The Rayleigh damping coefficients are selected as $\alpha_R = 0$ and $\beta_R = 0.1$, resulting in a damping coefficient $c = \beta_R k = 0.1\,\text{Ns/m}$. The example focuses on the first mode shape, which is associated with the complex eigenvalues $\lambda = -0.0191 \pm \text{i}0.6177$. The resulting SSM is expanded up to the 5-th order, and the displacement $x_2$ of the mass at the free end is considered.

### 6.4.1 Perturbed Backbone

We begin by computing the backbone curve of the system in nominal conditions (black lines in Fig. 6.3). Then, the adjoint method is used to evaluate the sensitivity of the backbone frequency $\Omega$ at a given reference amplitude $x = 1.5\,\text{m}$. The sensitivities are used to make predictions (cross markers in Fig. 6.3) of the backbone

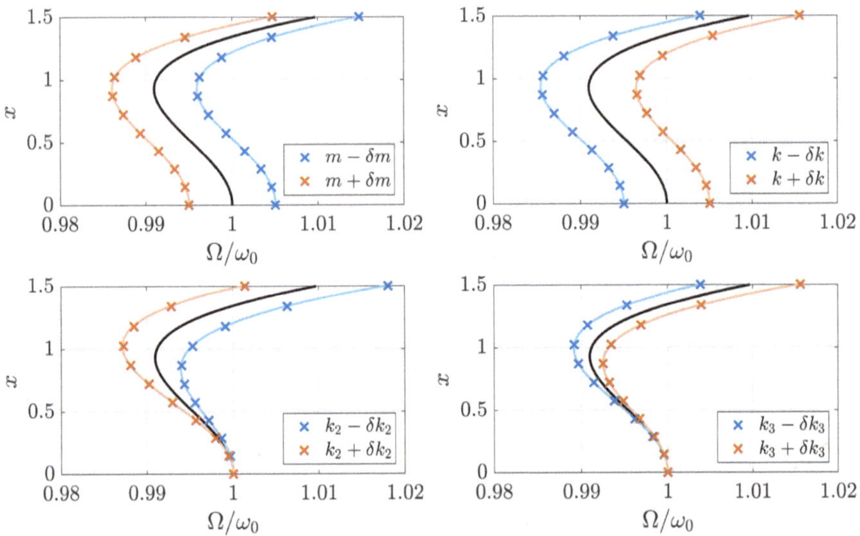


**Fig. 6.3** Backbone curve sensitivity analysis for the 2 DOF nonlinear oscillator. Starting from the backbone curve of the nominal system (black lines), the sensitivity analysis is used to predict the backbone curve of the perturbed system, as described in Eq. (6.21). The predictions (cross markers) are then validated by computing the backbone curve of the perturbed system (colored lines)

curve of the perturbed system:

$$\Omega(\mu \pm \delta\mu) = \Omega(\mu) \pm \frac{\mathrm{d}\Omega}{\mathrm{d}\mu}\delta\mu, \tag{6.21}$$

where the parameter perturbations are $\delta m = 0.01\,m$, $\delta k = 0.01\,k$, $\delta k_2 = 0.03\,k_2$, and $\delta k_3 = 0.03\,k_3$. The predictions are then validated by directly computing the backbone curve of the perturbed system (colored lines in Fig. 6.3).

On the one hand, the natural frequency is influenced only by perturbations in $m$ and $k$. On the other hand, changes in $k_2$ and $k_3$ become significant at higher amplitudes, thereby affecting the dynamic behavior of the response.

### *6.4.2 Perturbed Frequency Response Peak*

For the given reference amplitude $x = 1.5\,\mathrm{m}$, we first solve Eq. (5.3) to obtain the corresponding reduced amplitude $\rho$. This value is then used in Eq. (5.5) to determine the forcing level required to achieve the desired amplitude. Using this forcing level, we compute the sensitivity of the frequency response peak with respect to system parameters.

Compared to the previous case (Fig. 6.3), where only the frequency was affected, the peak frequency and amplitude are both influenced by parameter perturbations. These sensitivities allow us to predict the location of the perturbed system's frequency response peak, as shown by the cross markers in Fig. 6.4:

$$\begin{aligned} \Omega(\mu \pm \delta\mu) &= \Omega(\mu) \pm \frac{\mathrm{d}\Omega}{\mathrm{d}\mu}\delta\mu \\ x(\mu \pm \delta\mu) &= x(\mu) \pm \frac{\mathrm{d}x}{\mathrm{d}\mu}\delta\mu, \end{aligned} \tag{6.22}$$

where the parameter perturbations are $\delta m = 0.01\,m$, $\delta k = 0.01\,k$, $\delta k_2 = 0.02\,k_2$, and $\delta k_3 = 0.02\,k_3$. The predicted peaks are then validated by directly computing the frequency response of the perturbed system (colored lines in Fig. 6.4).

## 6.5 The Total Derivatives Handbook

The derivatives in Eqs. (6.3)–(6.9) depend on the coefficients of the autonomous and non-autonomous SSM coefficients. Here, we briefly report all these derivatives, which can be obtained by differentiating the expressions described in Chap. 2.

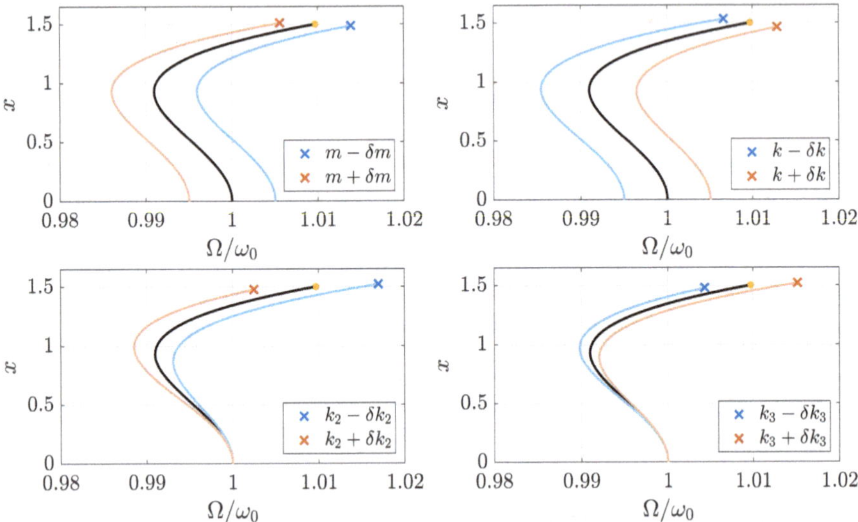


**Fig. 6.4** Frequency response peak sensitivity analysis for the 2 DOF nonlinear oscillator. Starting from the peak of the nominal response (yellow dot), the predicted peak locations (cross markers) under parameter perturbations are obtained via sensitivity analysis and validated against the backbone curves of the perturbed systems (colored lines)

We start from the derivatives of the master mode shape and eigenfrequency, which are solved together from the linear system

$$\begin{bmatrix} \mathbf{K}-\omega^2\mathbf{M} & -2\omega\mathbf{M}\boldsymbol{\varphi} \\ -2\omega\boldsymbol{\varphi}^T\mathbf{M} & 0 \end{bmatrix} \begin{bmatrix} \frac{\mathrm{d}\boldsymbol{\varphi}}{\mathrm{d}\mu} \\ \frac{\mathrm{d}\omega}{\mathrm{d}\mu} \end{bmatrix} = \begin{bmatrix} \left(\omega^2\frac{\mathrm{d}\mathbf{M}}{\mathrm{d}\mu}-\frac{\mathrm{d}\mathbf{K}}{\mathrm{d}\mu}\right)\boldsymbol{\varphi} \\ \omega\boldsymbol{\varphi}^T\frac{\mathrm{d}\mathbf{M}}{\mathrm{d}\mu}\boldsymbol{\varphi} \end{bmatrix}. \tag{6.23}$$

The sensitivity of the eigenvalue $\lambda$ is

$$\frac{\mathrm{d}\lambda}{\mathrm{d}\mu} = -\xi\frac{\mathrm{d}\omega}{\mathrm{d}\mu} - \omega\frac{\mathrm{d}\xi}{\mathrm{d}\mu} + \mathrm{i}\frac{\mathrm{d}\omega}{\mathrm{d}\mu}\sqrt{1-\xi^2} - \mathrm{i}\omega\frac{\xi}{\sqrt{1-\xi^2}}\frac{\mathrm{d}\xi}{\mathrm{d}\mu}, \tag{6.24}$$

**Damping Total Derivative**

The total derivatives of the damping terms depend on the selection of the Rayleigh damping coefficients $\alpha_R$ and $\beta_R$. A constant choice for $\alpha_R$ and $\beta_R$ leads to

$$\frac{\mathrm{d}\mathbf{C}}{\mathrm{d}\mu} = \alpha_R\frac{\mathrm{d}\mathbf{M}}{\mathrm{d}\mu} + \alpha_R\frac{\mathrm{d}\mathbf{M}}{\mathrm{d}\mu}$$

(continued)

$$\frac{\mathrm{d}\delta}{\mathrm{d}\mu} = 2\beta_R \omega \frac{\mathrm{d}\omega}{\mathrm{d}\mu}$$
$$\frac{\mathrm{d}\xi}{\mathrm{d}\mu} = \frac{\beta_R \omega^2 - \alpha_R}{2\omega^2} \frac{\mathrm{d}\omega}{\mathrm{d}\mu}.$$

Otherwise, if a constant Q-factor is selected, the derivatives of the damping terms are:

$$\frac{\mathrm{d}\mathbf{C}}{\mathrm{d}\mu} = \frac{\partial \mathbf{C}}{\partial \mu} + \frac{\partial \mathbf{C}}{\partial \omega} \frac{\mathrm{d}\omega}{\mathrm{d}\mu}$$
$$\frac{\mathrm{d}\delta}{\mathrm{d}\mu} = 2\xi \frac{\mathrm{d}\omega}{\mathrm{d}\mu}$$
$$\frac{\mathrm{d}\xi}{\mathrm{d}\mu} = 0.$$

In this work, we focus on the two easy possibilities to achieve a constant Q-factor, and this choice affects the derivative of the damping matrix $\mathbf{C}$:

$$\text{if} \quad \alpha_R = 0 \quad \text{and} \quad \beta_R = \frac{1}{\omega Q} \implies \frac{\mathrm{d}\mathbf{C}}{\mathrm{d}\mu} = \frac{1}{\omega Q} \frac{\mathrm{d}\mathbf{K}}{\mathrm{d}\mu} - \frac{1}{\omega^2 Q} \frac{\mathrm{d}\omega}{\mathrm{d}\mu} \mathbf{K}$$
$$\text{if} \quad \alpha_R = \frac{\omega}{Q} \quad \text{and} \quad \beta_R = 0 \implies \frac{\mathrm{d}\mathbf{C}}{\mathrm{d}\mu} = \frac{\omega}{Q} \frac{\mathrm{d}\mathbf{M}}{\mathrm{d}\mu} + \frac{1}{Q} \frac{\mathrm{d}\omega}{\mathrm{d}\mu} \mathbf{M}.$$

The leading-order coefficients of the autonomous SSM directly depend on the master mode shape and eigenfrequency. Therefore, the sensitivities are simply

$$\frac{\mathrm{d}R^1_{10}}{\mathrm{d}\mu} = \frac{\mathrm{d}\bar{R}^2_{01}}{\mathrm{d}\mu} = \frac{\mathrm{d}\lambda}{\mathrm{d}\mu} \tag{6.25a}$$

$$\frac{\mathrm{d}\boldsymbol{w}_{10}}{\mathrm{d}\mu} = \frac{\mathrm{d}\bar{\boldsymbol{w}}_{01}}{\mathrm{d}\mu} = \frac{\mathrm{d}\boldsymbol{\varphi}}{\mathrm{d}\mu} \tag{6.25b}$$

$$\frac{\mathrm{d}\dot{\boldsymbol{w}}_{10}}{\mathrm{d}\mu} = \frac{\mathrm{d}\dot{\bar{\boldsymbol{w}}}_{01}}{\mathrm{d}\mu} = \frac{\mathrm{d}\lambda}{\mathrm{d}\mu}\boldsymbol{\varphi} + \lambda \frac{\mathrm{d}\boldsymbol{\varphi}}{\mathrm{d}\mu}. \tag{6.25c}$$

We now list the sensitivities of the autonomous SSM coefficients, starting from the eigenvalue coefficient $\Lambda_{\boldsymbol{m}}$:

$$\frac{\mathrm{d}\Lambda_{\boldsymbol{m}}}{\mathrm{d}\mu} = \boldsymbol{m} \cdot \frac{\mathrm{d}\boldsymbol{\Lambda}}{\mathrm{d}\mu} = m_1 \frac{\mathrm{d}\lambda}{\mathrm{d}\mu} + m_2 \frac{\mathrm{d}\bar{\lambda}}{\mathrm{d}\mu}. \tag{6.26}$$

Using the tensor notation, the sensitivity of the nonlinear force contribution is

$$
\begin{aligned}
\frac{\mathrm{d}f^i_{\boldsymbol{m}}}{\mathrm{d}\mu} = & \sum_{\substack{\boldsymbol{u},\boldsymbol{k}\in\mathbb{N}^2 \\ \boldsymbol{u}+\boldsymbol{k}=\boldsymbol{m}}} \left[ \frac{\mathrm{d}T_2^{ijk}}{\mathrm{d}\mu} w^j_{\boldsymbol{u}} w^k_{\boldsymbol{k}} + T_2^{ijk} \left( \frac{\mathrm{d}w^j_{\boldsymbol{u}}}{\mathrm{d}\mu} w^k_{\boldsymbol{k}} + w^j_{\boldsymbol{u}} \frac{\mathrm{d}w^k_{\boldsymbol{k}}}{\mathrm{d}\mu} \right) \right] \\
& + \sum_{\substack{\boldsymbol{u},\boldsymbol{k},\boldsymbol{l}\in\mathbb{N}^2 \\ \boldsymbol{u}+\boldsymbol{k}+\boldsymbol{l}=\boldsymbol{m}}} \left[ \frac{\mathrm{d}T_3^{ijkl}}{\mathrm{d}\mu} w^j_{\boldsymbol{u}} w^k_{\boldsymbol{k}} w^l_{\boldsymbol{l}} \right. \\
& \left. + T_3^{ijkl} \left( \frac{\mathrm{d}w^j_{\boldsymbol{u}}}{\mathrm{d}\mu} w^k_{\boldsymbol{k}} w^l_{\boldsymbol{l}} + w^j_{\boldsymbol{u}} \frac{\mathrm{d}w^k_{\boldsymbol{k}}}{\mathrm{d}\mu} w^l_{\boldsymbol{l}} + w^j_{\boldsymbol{u}} w^k_{\boldsymbol{k}} \frac{\mathrm{d}w^l_{\boldsymbol{l}}}{\mathrm{d}\mu} \right) \right].
\end{aligned}
\tag{6.27}
$$

The derivatives of the lower-order manifold contributions are

$$
\frac{\mathrm{d}\boldsymbol{V}_{\boldsymbol{m}}}{\mathrm{d}\mu} = \sum_{j=1}^{2} \sum_{\substack{\boldsymbol{u},\boldsymbol{k}\in\mathbb{N}^2 \\ 1<u,k<m \\ \boldsymbol{u}+\boldsymbol{k}-\boldsymbol{e}_j=\boldsymbol{m}}} u_j \left( \frac{\mathrm{d}\boldsymbol{w}_{\boldsymbol{u}}}{\mathrm{d}\mu} R^j_{\boldsymbol{k}} + \boldsymbol{w}_{\boldsymbol{u}} \frac{\mathrm{d}R^j_{\boldsymbol{k}}}{\mathrm{d}\mu} \right) \tag{6.28}
$$

$$
\frac{\mathrm{d}\dot{\boldsymbol{V}}_{\boldsymbol{m}}}{\mathrm{d}\mu} = \sum_{j=1}^{2} \sum_{\substack{\boldsymbol{u},\boldsymbol{k}\in\mathbb{N}^2 \\ 1<u,k<m \\ \boldsymbol{u}+\boldsymbol{k}-\boldsymbol{e}_j=\boldsymbol{m}}} u_j \left( \frac{\mathrm{d}\dot{\boldsymbol{w}}_{\boldsymbol{u}}}{\mathrm{d}\mu} R^j_{\boldsymbol{k}} + \dot{\boldsymbol{w}}_{\boldsymbol{u}} \frac{\mathrm{d}R^j_{\boldsymbol{k}}}{\mathrm{d}\mu} \right), \tag{6.29}
$$

which contribute to the sensitivity of the vector $\boldsymbol{C}_{\boldsymbol{m}}$:

$$
\begin{aligned}
\frac{\mathrm{d}\boldsymbol{C}_{\boldsymbol{m}}}{\mathrm{d}\mu} = & -\frac{\mathrm{d}\mathbf{M}}{\mathrm{d}\mu} \dot{\boldsymbol{V}}_{\boldsymbol{m}} - \mathbf{M} \frac{\mathrm{d}\dot{\boldsymbol{V}}_{\boldsymbol{m}}}{\mathrm{d}\mu} - (\Lambda_{\boldsymbol{m}} \mathbf{M} + \mathbf{C}) \frac{\mathrm{d}\boldsymbol{V}_{\boldsymbol{m}}}{\mathrm{d}\mu} \\
& - \left( \frac{\mathrm{d}\Lambda_{\boldsymbol{m}}}{\mathrm{d}\mu} \mathbf{M} + \Lambda_{\boldsymbol{m}} \frac{\mathrm{d}\mathbf{M}}{\mathrm{d}\mu} + \frac{\mathrm{d}\mathbf{C}}{\mathrm{d}\mu} \right) \boldsymbol{V}_{\boldsymbol{m}} - \frac{\mathrm{d}\boldsymbol{f}_{\boldsymbol{m}}}{\mathrm{d}\mu}.
\end{aligned}
\tag{6.30}
$$

The sensitivity of the reduced dynamic coefficients is

$$
\begin{aligned}
\frac{\mathrm{d}R^j_{\boldsymbol{m}}}{\mathrm{d}\mu} = & \frac{\frac{\mathrm{d}\boldsymbol{\varphi}^T}{\mathrm{d}\mu} \boldsymbol{C}_{\boldsymbol{m}} + \boldsymbol{\varphi}^T \frac{\mathrm{d}\boldsymbol{C}_{\boldsymbol{m}}}{\mathrm{d}\mu}}{\Lambda_{\boldsymbol{m}} + \lambda_j + \delta} \\
& - R^j_{\boldsymbol{m}} \frac{\frac{\mathrm{d}\Lambda_{\boldsymbol{m}}}{\mathrm{d}\mu} + \frac{\mathrm{d}\lambda_j}{\mathrm{d}\mu} + \frac{\mathrm{d}\delta}{\mathrm{d}\mu}}{\Lambda_{\boldsymbol{m}} + \lambda_j + \delta}, \quad j = 1, 2,
\end{aligned}
\tag{6.31}
$$

if $\boldsymbol{m}$ and $j$ satisfy the near-resonance condition, 0 otherwise.

The sensitivity of the manifold coefficients $\boldsymbol{w}_m$ is computed by differentiating and solving the cohomological equation (2.21):

$$\mathbf{L}_m \frac{\mathrm{d}\boldsymbol{w}_m}{\mathrm{d}\mu} = \frac{\mathrm{d}\boldsymbol{h}_m}{\mathrm{d}\mu} - \frac{\mathrm{d}\mathbf{L}_m}{\mathrm{d}\mu}\boldsymbol{w}_m \tag{6.32}$$

where

$$\frac{\mathrm{d}\mathbf{L}_m}{\mathrm{d}\mu} = \frac{\mathrm{d}\mathbf{K}}{\mathrm{d}\mu} + \Lambda_m \frac{\mathrm{d}\mathbf{C}}{\mathrm{d}\mu} + \Lambda_m^2 \frac{\mathrm{d}\mathbf{M}}{\mathrm{d}\mu} + (\mathbf{C} + 2\Lambda_m \mathbf{M}) \frac{\mathrm{d}\Lambda_m}{\mathrm{d}\mu}, \tag{6.33}$$

$$\frac{\mathrm{d}\boldsymbol{h}_m}{\mathrm{d}\mu} = \frac{\mathrm{d}\boldsymbol{C}_m}{\mathrm{d}\mu} + \sum_{j=1}^{2} \left( \frac{\mathrm{d}\boldsymbol{D}_m^j}{\mathrm{d}\mu} R_m^j + \boldsymbol{D}_m^j \frac{\mathrm{d}R_m^j}{\mathrm{d}\mu} \right), \tag{6.34}$$

$$\begin{aligned} \frac{\mathrm{d}\boldsymbol{D}_m^j}{\mathrm{d}\mu} = &- \left[ (\Lambda_m + \lambda_j) \mathbf{M} + \mathbf{C} \right] \frac{\mathrm{d}\boldsymbol{\varphi}}{\mathrm{d}\mu} \\ &- \left[ \left( \frac{\mathrm{d}\Lambda_m}{\mathrm{d}\mu} + \frac{\mathrm{d}\lambda_j}{\mathrm{d}\mu} \right) \mathbf{M} \right. \\ &\left. + (\Lambda_m + \lambda_j) \frac{\mathrm{d}\mathbf{M}}{\mathrm{d}\mu} + \frac{\mathrm{d}\mathbf{C}}{\mathrm{d}\mu} \right] \boldsymbol{\varphi}, \quad j = 1, 2. \end{aligned} \tag{6.35}$$

The sensitivity of the manifold velocity coefficients $\dot{\boldsymbol{w}}_m$ are then computed from

$$\begin{aligned} \frac{\mathrm{d}\dot{\boldsymbol{w}}_m}{\mathrm{d}\mu} = &\frac{\mathrm{d}\Lambda_m}{\mathrm{d}\mu} \boldsymbol{w}_m + \Lambda_m \frac{\mathrm{d}\boldsymbol{w}_m}{\mathrm{d}\mu} + \frac{\mathrm{d}\boldsymbol{V}_m}{\mathrm{d}\mu} \\ &+ \sum_{j=1}^{2} \left( \frac{\mathrm{d}R_m^j}{\mathrm{d}\mu} \boldsymbol{\varphi} + R_m^j \frac{\mathrm{d}\boldsymbol{\varphi}}{\mathrm{d}\mu} \right). \end{aligned} \tag{6.36}$$

Finally, the sensitivity of the non-autonomous reduced dynamic coefficients $S_0$ is

$$\frac{\mathrm{d}S_0}{\mathrm{d}\mu} = - \frac{\frac{\mathrm{d}\boldsymbol{\varphi}^T}{\mathrm{d}\mu} \boldsymbol{g}_0}{\Lambda_0 + \lambda + \delta} - S_0 \frac{\frac{\mathrm{d}\Lambda_0}{\mathrm{d}\mu} + \frac{\mathrm{d}\lambda}{\mathrm{d}\mu} + \frac{\mathrm{d}\delta}{\mathrm{d}\mu}}{\Lambda_0 + \lambda + \delta}, \tag{6.37}$$

where

$$\frac{\mathrm{d}\Lambda_0}{\mathrm{d}\mu} = \mathrm{i} \frac{\mathrm{d}\Omega}{\mathrm{d}\mu} = \mathrm{i}\, \mathrm{Im} \left\{ \frac{\mathrm{d}\lambda}{\mathrm{d}\mu} \right\}. \tag{6.38}$$

Similarly to the SSM computation steps, Eqs. (6.26)–(6.36) are computed for every multi-index, starting from those at the lowest order. These computations can be carried out alongside the primary SSM analysis.

## 6.6 The Partial Derivatives Handbook

The adjoint sensitivities described in Sect. 6.3 depend on the autonomous and non-autonomous SSM coefficients. In this section, we list all the partial derivatives required in Eqs. (6.11)–(6.19).

Before doing so, we note that the partial derivative of a vector $\boldsymbol{a}$ with respect to another vector $\boldsymbol{b}$ leads to a matrix that, in general, can be dense. To avoid storing this full matrix, which would be unfeasible even for relatively small systems, we instead store the product $\boldsymbol{v}^T \frac{\partial \boldsymbol{a}}{\partial \boldsymbol{b}}$, where $\boldsymbol{v}$ is the vector that multiplies $\boldsymbol{a}$ is the sensitivity expression. If $\boldsymbol{a}$ appears in multiple expressions, multiplied by different vectors, say $\boldsymbol{u}$ and $\boldsymbol{v}$, then the products $\boldsymbol{u}^T \frac{\partial \boldsymbol{a}}{\partial \boldsymbol{b}}$ and $\boldsymbol{v}^T \frac{\partial \boldsymbol{a}}{\partial \boldsymbol{b}}$ must be evaluated separately, as the derivative $\frac{\partial \boldsymbol{a}}{\partial \boldsymbol{b}}$ cannot be stored.

### 6.6.1 *Partial Derivatives with Respect to the Reduced Amplitude*

The only terms that explicitly depend on the reduced amplitude $\rho$ are $\Omega$, $x$, and $a$. Their partial derivatives with respect to $\rho$ are

$$\frac{\partial \Omega}{\partial \rho} = \frac{1}{2}\mathrm{i} \sum_{m>1} \left(\gamma_m^2 - \gamma_m^1\right)(m-1)\rho^{m-2} \tag{6.39}$$

$$\frac{\partial x}{\partial \rho} = \frac{1}{x N_t} \sum_{k=1}^{N_t} x_k^i \sum_{\boldsymbol{m} \in \mathbb{N}^2} w_{\boldsymbol{m}}^i m \rho^{m-1} \tilde{\boldsymbol{p}}_k^{\boldsymbol{m}} \tag{6.40}$$

$$\frac{\partial a}{\partial \rho} = \frac{1}{2} \sum_{m \in \mathbb{N}} \left(\gamma_m^2 + \gamma_m^1\right) m \rho^{m-1}. \tag{6.41}$$

### 6.6.2 *Partial Derivatives with Respect to the Manifold Coefficients*

Before delving into the partial derivatives with respect to the manifold coefficients $\boldsymbol{w}_{\boldsymbol{m}}$, it is essential to note that the autonomous SSM coefficients at a given order $q$ are independent of the manifold coefficients at higher orders. As a result, the partial derivative of any term at order $q$ with respect to a manifold coefficient at order $m$ is zero for all $m \geq q$.

Using tensor notation, the partial derivative of the nonlinear force contribution is

$$\begin{aligned}\frac{\partial f_{\boldsymbol{q}}^i}{\partial w_{\boldsymbol{m}}^p} &= \sum_{\substack{\boldsymbol{u},\boldsymbol{k}\in\mathbb{N}^2\\ \boldsymbol{u}+\boldsymbol{k}=\boldsymbol{q}}} T_2^{ijk}\left(\frac{\partial w_{\boldsymbol{u}}^j}{\partial w_{\boldsymbol{m}}^p} w_{\boldsymbol{k}}^k + w_{\boldsymbol{u}}^j \frac{\partial w_{\boldsymbol{k}}^k}{\partial w_{\boldsymbol{m}}^p}\right)\\ &+ \sum_{\substack{\boldsymbol{u},\boldsymbol{k},\boldsymbol{l}\in\mathbb{N}^2\\ \boldsymbol{u}+\boldsymbol{k}+\boldsymbol{l}=\boldsymbol{q}}} T_3^{ijkl}\left(\frac{\partial w_{\boldsymbol{u}}^j}{\partial w_{\boldsymbol{m}}^p} w_{\boldsymbol{k}}^k w_{\boldsymbol{l}}^l + w_{\boldsymbol{u}}^j \frac{\partial w_{\boldsymbol{k}}^k}{\partial w_{\boldsymbol{m}}^p} w_{\boldsymbol{l}}^l + w_{\boldsymbol{u}}^j w_{\boldsymbol{k}}^k \frac{\partial w_{\boldsymbol{l}}^l}{\partial w_{\boldsymbol{m}}^p}\right),\end{aligned} \tag{6.42}$$

where

$$\frac{\partial w_{\boldsymbol{u}}^j}{\partial w_{\boldsymbol{m}}^p} = \begin{cases} 1 & \text{if} \quad \boldsymbol{u} = \boldsymbol{m} \quad \text{and} \quad p = j \\ 0 & \text{otherwise.} \end{cases} \tag{6.43}$$

The same condition applies to all the other partial derivatives in Eq. (6.42).
The partial derivatives of vectors $\boldsymbol{V}_{\boldsymbol{q}}$ and $\dot{\boldsymbol{V}}_{\boldsymbol{q}}$ are

$$\boldsymbol{v}^T \frac{\partial \boldsymbol{V}_{\boldsymbol{q}}}{\partial \boldsymbol{w}_{\boldsymbol{m}}} = \sum_{j=1}^{2} \sum_{\substack{\boldsymbol{u},\boldsymbol{k}\in\mathbb{N}^2\\ 1<u,k<q\\ \boldsymbol{u}+\boldsymbol{k}-\boldsymbol{e}_j=\boldsymbol{q}}} u_j \left(\boldsymbol{v}^T \frac{\partial \boldsymbol{w}_{\boldsymbol{u}}}{\partial \boldsymbol{w}_{\boldsymbol{m}}} R_{\boldsymbol{k}}^j + (\boldsymbol{v}^T \boldsymbol{w}_{\boldsymbol{u}}) \frac{\partial R_{\boldsymbol{k}}^j}{\partial \boldsymbol{w}_{\boldsymbol{m}}}\right) \tag{6.44}$$

$$\boldsymbol{v}^T \frac{\partial \dot{\boldsymbol{V}}_{\boldsymbol{q}}}{\partial \boldsymbol{w}_{\boldsymbol{m}}} = \sum_{j=1}^{2} \sum_{\substack{\boldsymbol{u},\boldsymbol{k}\in\mathbb{N}^2\\ 1<u,k<q\\ \boldsymbol{u}+\boldsymbol{k}-\boldsymbol{e}_j=\boldsymbol{q}}} u_j \left(\boldsymbol{v}^T \frac{\partial \dot{\boldsymbol{w}}_{\boldsymbol{u}}}{\partial \boldsymbol{w}_{\boldsymbol{m}}} R_{\boldsymbol{k}}^j + (\boldsymbol{v}^T \dot{\boldsymbol{w}}_{\boldsymbol{u}}) \frac{\partial R_{\boldsymbol{k}}^j}{\partial \boldsymbol{w}_{\boldsymbol{m}}}\right), \tag{6.45}$$

where $\boldsymbol{v}$ is a generic vector that multiplies $\boldsymbol{V}_{\boldsymbol{q}}$ and $\dot{\boldsymbol{V}}_{\boldsymbol{q}}$ in the Lagrangian function. The products $\boldsymbol{v}^T \boldsymbol{w}_{\boldsymbol{u}}$ and $\boldsymbol{v}^T \dot{\boldsymbol{w}}_{\boldsymbol{u}}$ are scalar quantities, and the term $\boldsymbol{v}^T \frac{\partial \boldsymbol{w}_{\boldsymbol{u}}}{\partial \boldsymbol{w}_{\boldsymbol{m}}}$ is equal to $\boldsymbol{v}^T$ if $\boldsymbol{u} = \boldsymbol{m}$, otherwise it is equal to a null row vector.

The partial derivative of vector $\boldsymbol{C}_{\boldsymbol{q}}$ is

$$\boldsymbol{v}^T \frac{\partial \boldsymbol{C}_{\boldsymbol{q}}}{\partial \boldsymbol{w}_{\boldsymbol{m}}} = -\boldsymbol{v}^T \mathbf{M} \frac{\partial \dot{\boldsymbol{V}}_{\boldsymbol{q}}}{\partial \boldsymbol{w}_{\boldsymbol{m}}} - \boldsymbol{v}^T \frac{\partial \boldsymbol{f}_{\boldsymbol{q}}}{\partial \boldsymbol{w}_{\boldsymbol{m}}} - \boldsymbol{v}^T \left(\Lambda_{\boldsymbol{q}} \mathbf{M} + \mathbf{C}\right) \frac{\partial \boldsymbol{V}_{\boldsymbol{q}}}{\partial \boldsymbol{w}_{\boldsymbol{m}}}. \tag{6.46}$$

The partial derivative of the reduced dynamics coefficient is

$$\frac{\partial R_{\boldsymbol{q}}^j}{\partial \boldsymbol{w}_{\boldsymbol{m}}} = \frac{\boldsymbol{\varphi}^T \frac{\partial \boldsymbol{C}_{\boldsymbol{q}}}{\partial \boldsymbol{w}_{\boldsymbol{m}}}}{\Lambda_{\boldsymbol{q}} + \lambda_j + \delta}, \quad j = 1, 2, \tag{6.47}$$

where $\boldsymbol{\varphi}^T \frac{\partial \boldsymbol{C}_{\boldsymbol{q}}}{\partial \boldsymbol{w}_{\boldsymbol{m}}}$ is computed using Eq. (6.46) by setting $\boldsymbol{v} = \boldsymbol{\varphi}$.

The partial derivative of the right-hand side of the cohomological equation is

$$\boldsymbol{v}^T \frac{\partial \boldsymbol{h}_{\boldsymbol{q}}}{\partial \boldsymbol{w}_{\boldsymbol{m}}} = -\boldsymbol{v}^T \frac{\partial \boldsymbol{C}_{\boldsymbol{q}}}{\partial \boldsymbol{w}_{\boldsymbol{m}}} - \sum_{j=1}^{2} (\boldsymbol{v}^T \boldsymbol{D}_{\boldsymbol{q}}^j) \frac{\partial R_{\boldsymbol{q}}^j}{\partial \boldsymbol{w}_{\boldsymbol{m}}}, \tag{6.48}$$

where the product $\boldsymbol{v}^T \boldsymbol{D}_{\boldsymbol{q}}^j$ is a scalar quantity, and the vector $\boldsymbol{D}_{\boldsymbol{q}}^j$ does not depend on the manifold coefficients.

The partial derivative of the manifold velocity coefficient is

$$\boldsymbol{v}^T \frac{\partial \dot{\boldsymbol{w}}_{\boldsymbol{q}}}{\partial \boldsymbol{w}_{\boldsymbol{m}}} = \Lambda_{\boldsymbol{q}} \boldsymbol{v}^T \frac{\partial \boldsymbol{w}_{\boldsymbol{q}}}{\partial \boldsymbol{w}_{\boldsymbol{m}}} + (\boldsymbol{v}^T \boldsymbol{\varphi}) \sum_{j=1}^{2} \frac{\partial R_{\boldsymbol{q}}^j}{\partial \boldsymbol{w}_{\boldsymbol{m}}} + \boldsymbol{v}^T \frac{\partial \boldsymbol{V}_{\boldsymbol{q}}}{\partial \boldsymbol{w}_{\boldsymbol{m}}}, \tag{6.49}$$

where the product $\boldsymbol{v}^T \boldsymbol{\varphi}$ is a scalar quantity.

The non-autonomous SSM coefficient $S_0$ does not depend on the autonomous manifold coefficients, hence its partial derivative with respect to $\boldsymbol{w}_{\boldsymbol{m}}$ is zero.

Finally, the partial derivative of $a$, $\Omega$, and $x$ are

$$\frac{\partial a}{\partial \boldsymbol{w}_{\boldsymbol{m}}} = \frac{1}{2} \sum_{q>m} \left( \frac{\partial \gamma_q^2}{\partial \boldsymbol{w}_{\boldsymbol{m}}} + \frac{\partial \gamma_q^1}{\partial \boldsymbol{w}_{\boldsymbol{m}}} \right) \rho^q. \tag{6.50}$$

$$\frac{\partial \Omega}{\partial \boldsymbol{w}_{\boldsymbol{m}}} = \frac{1}{2} \mathrm{i} \sum_{q>m} \left( \frac{\partial \gamma_q^2}{\partial \boldsymbol{w}_{\boldsymbol{m}}} - \frac{\partial \gamma_q^1}{\partial \boldsymbol{w}_{\boldsymbol{m}}} \right) \rho^{q-1}. \tag{6.51}$$

$$\frac{\partial x}{\partial w_{\boldsymbol{m}}^j} = \begin{cases} \dfrac{1}{x N_t} \displaystyle\sum_{k=1}^{N_t} x_k^i \boldsymbol{p}_k^{\boldsymbol{m}} & \text{if} \quad j = i \\ 0 & \text{otherwise.} \end{cases} \tag{6.52}$$

### 6.6.3 *Partial Derivatives with Respect to the Master Mode Shape*

Using tensor notation, the partial derivative of the nonlinear force contribution is

$$\begin{aligned} \frac{\partial f_{\boldsymbol{m}}^i}{\partial \varphi^p} = & \sum_{\substack{\boldsymbol{u},\boldsymbol{k} \in \mathbb{N}^2 \\ \boldsymbol{u}+\boldsymbol{k}=\boldsymbol{m}}} T_2^{ijk} \left( \frac{\partial w_{\boldsymbol{u}}^j}{\partial \varphi^p} w_{\boldsymbol{k}}^k + w_{\boldsymbol{u}}^j \frac{\partial w_{\boldsymbol{k}}^k}{\partial \varphi^p} \right) \\ & + \sum_{\substack{\boldsymbol{u},\boldsymbol{k},\boldsymbol{l} \in \mathbb{N}^2 \\ \boldsymbol{u}+\boldsymbol{k}+\boldsymbol{l}=\boldsymbol{m}}} T_3^{ijkl} \left( \frac{\partial w_{\boldsymbol{u}}^j}{\partial \varphi^p} w_{\boldsymbol{k}}^k w_{\boldsymbol{l}}^l + w_{\boldsymbol{u}}^j \frac{\partial w_{\boldsymbol{k}}^k}{\partial \varphi^p} w_{\boldsymbol{l}}^l + w_{\boldsymbol{u}}^j w_{\boldsymbol{k}}^k \frac{\partial w_{\boldsymbol{l}}^l}{\partial \varphi^p} \right), \end{aligned} \tag{6.53}$$

where the partial derivative $\frac{\partial w_{\boldsymbol{u}}^j}{\partial \varphi^p}$ is defined as follow:

$$\frac{\partial w_{\boldsymbol{u}}^j}{\partial \varphi^p} = \begin{cases} 1 & \text{if} \quad |\boldsymbol{u}|_1 = 1 \quad \text{and} \quad p = j \\ 0 & \text{otherwise} \end{cases} \tag{6.54}$$

The same applies to all the other partial derivatives in Eq. (6.53). As before, the product $\boldsymbol{v}^T \frac{\partial \boldsymbol{f}_m}{\partial \varphi}$ is computed and stored instead of the full derivative matrix.

The partial derivatives of vectors $\boldsymbol{V}_{\boldsymbol{m}}$ and $\dot{\boldsymbol{V}}_{\boldsymbol{m}}$ are

$$\boldsymbol{v}^T \frac{\partial \boldsymbol{V}_{\boldsymbol{m}}}{\partial \varphi} = \sum_{j=1}^{2} \sum_{\substack{\boldsymbol{u},\boldsymbol{k} \in \mathbb{N}^2 \\ 1<u,k<m \\ \boldsymbol{u}+\boldsymbol{k}-\boldsymbol{e}_j=\boldsymbol{m}}} u_j (\boldsymbol{v}^T \boldsymbol{w}_{\boldsymbol{u}}) \frac{\partial R_{\boldsymbol{k}}^j}{\partial \varphi} \tag{6.55}$$

$$\boldsymbol{v}^T \frac{\partial \dot{\boldsymbol{V}}_{\boldsymbol{m}}}{\partial \varphi} = \sum_{j=1}^{2} \sum_{\substack{\boldsymbol{u},\boldsymbol{k} \in \mathbb{N}^2 \\ 1<u,k<m \\ \boldsymbol{u}+\boldsymbol{k}-\boldsymbol{e}_j=\boldsymbol{m}}} u_j \left( \boldsymbol{v}^T \frac{\partial \dot{\boldsymbol{w}}_{\boldsymbol{u}}}{\partial \varphi} R_{\boldsymbol{k}}^j + (\boldsymbol{v}^T \dot{\boldsymbol{w}}_{\boldsymbol{u}}) \frac{\partial R_{\boldsymbol{k}}^j}{\partial \varphi} \right), \tag{6.56}$$

where the products $\boldsymbol{v}^T \boldsymbol{w}_{\boldsymbol{u}}$ and $\boldsymbol{v}^T \dot{\boldsymbol{w}}_{\boldsymbol{u}}$ are scalar quantities.

The partial derivative of vector $\boldsymbol{C}_{\boldsymbol{m}}$ is

$$\boldsymbol{v}^T \frac{\partial \boldsymbol{C}_{\boldsymbol{m}}}{\partial \varphi} = -\boldsymbol{v}^T \mathbf{M} \frac{\partial \dot{\boldsymbol{V}}_{\boldsymbol{m}}}{\partial \varphi} - \boldsymbol{v}^T \frac{\partial \boldsymbol{f}_{\boldsymbol{m}}}{\partial \varphi} - \boldsymbol{v}^T \left( \Lambda_{\boldsymbol{m}} \mathbf{M} + \mathbf{C} \right) \frac{\partial \boldsymbol{V}_{\boldsymbol{q}}}{\partial \varphi}. \tag{6.57}$$

The partial derivative of the reduced dynamics coefficient is

$$\frac{\partial R_{\boldsymbol{m}}^j}{\partial \varphi} = \frac{\boldsymbol{C}_{\boldsymbol{m}}^T + \boldsymbol{\varphi}^T \frac{\partial \boldsymbol{C}_{\boldsymbol{m}}}{\partial \varphi}}{\Lambda_{\boldsymbol{q}} + \lambda_j + \delta}, \quad j = 1, 2, \tag{6.58}$$

where $\boldsymbol{\varphi}^T \frac{\partial \boldsymbol{C}_{\boldsymbol{m}}}{\partial \varphi}$ is computed using Eq. (6.57) by setting $\boldsymbol{v} = \boldsymbol{\varphi}$.

The partial derivative of vector $\boldsymbol{D}_{\boldsymbol{m}}^j$ is

$$\boldsymbol{v}^T \frac{\partial \boldsymbol{D}_{\boldsymbol{m}}^j}{\partial \varphi} = -\boldsymbol{v}^T \left[ \left( \Lambda_{\boldsymbol{m}} + \lambda_j \right) \mathbf{M} + \mathbf{C} \right], \quad j = 1, 2. \tag{6.59}$$

The partial derivative of the right-hand side of the cohomological equation is

$$\boldsymbol{v}^T \frac{\partial \boldsymbol{h}_{\boldsymbol{m}}}{\partial \varphi} = -\boldsymbol{v}^T \frac{\partial \boldsymbol{C}_{\boldsymbol{m}}}{\partial \varphi} - \sum_{j=1}^{2} (\boldsymbol{v}^T \boldsymbol{D}_{\boldsymbol{m}}^j) \frac{\partial R_{\boldsymbol{m}}^j}{\partial \varphi} - \sum_{j=1}^{2} \boldsymbol{v}^T \frac{\partial \boldsymbol{D}_{\boldsymbol{m}}^j}{\partial \varphi} R_{\boldsymbol{m}}^j \tag{6.60}$$

where the product $\boldsymbol{v}^T \boldsymbol{D}_{\boldsymbol{m}}^j$ is a scalar quantity.

The partial derivative of the manifold velocity coefficient is

$$\boldsymbol{v}^T \frac{\partial \dot{\boldsymbol{w}}_{\boldsymbol{m}}}{\partial \boldsymbol{\varphi}} = (\boldsymbol{v}^T \boldsymbol{\varphi}) \sum_{j=1}^{2} \frac{\partial R_{\boldsymbol{m}}^j}{\partial \boldsymbol{\varphi}} + \boldsymbol{v}^T \sum_{j=1}^{2} R_{\boldsymbol{m}}^j + \boldsymbol{v}^T \frac{\partial \boldsymbol{V}_{\boldsymbol{m}}}{\partial \boldsymbol{\varphi}} \tag{6.61}$$

where the product $\boldsymbol{v}^T \boldsymbol{\varphi}$ is a scalar quantity.

The partial derivative of the non-autonomous SSM coefficient $S_0$ is

$$\frac{\partial S_0}{\partial \boldsymbol{\varphi}} = \frac{-\boldsymbol{g}_0^T}{\Lambda_0 + \lambda + \delta}. \tag{6.62}$$

Finally, the partial derivatives of $a$, $\Omega$, and $x$ are

$$\frac{\partial a}{\partial \boldsymbol{\varphi}} = \frac{1}{2} \sum_{m \in \mathbb{N}} \left( \frac{\partial \gamma_m^2}{\partial \boldsymbol{\varphi}} + \frac{\partial \gamma_m^1}{\partial \boldsymbol{\varphi}} \right) \rho^m \tag{6.63}$$

$$\frac{\partial \Omega}{\partial \boldsymbol{\varphi}} = \frac{1}{2} \mathrm{i} \sum_{m > 1} \left( \frac{\partial \gamma_m^2}{\partial \boldsymbol{\varphi}} - \frac{\partial \gamma_m^1}{\partial \boldsymbol{\varphi}} \right) \rho^{m-1} \tag{6.64}$$

$$\frac{\partial x}{\partial \varphi^j} = \begin{cases} \dfrac{1}{x N_t} \displaystyle\sum_{k=1}^{N_t} x_k^i \left( \boldsymbol{p}_k^{10} + \boldsymbol{p}_k^{01} \right) & \text{if} \quad j = i \\ 0 & \text{otherwise.} \end{cases} \tag{6.65}$$

### 6.6.4 Partial Derivatives with Respect to the Master Eigenfrequency

Here, the partial derivatives are taken with respect to the eigenfrequency $\omega$. Since $\omega$ is a scalar, there is no need to multiply the partial derivatives by $\boldsymbol{v}^T$ as done in the previous cases.

The partial derivative of the eigenvalue coefficient is

$$\frac{\partial \Lambda_{\boldsymbol{m}}}{\partial \omega} = \boldsymbol{m} \cdot \frac{\partial \Lambda}{\partial \omega}, \tag{6.66}$$

where

$$\frac{\partial \Lambda}{\partial \omega} = \begin{bmatrix} \frac{\partial \lambda}{\partial \omega} \\ \frac{\partial \bar{\lambda}}{\partial \omega} \end{bmatrix} \tag{6.67}$$

$$\frac{\partial \lambda}{\partial \omega} = -\xi - \omega \frac{\partial \xi}{\partial \omega} + \mathrm{i}\sqrt{1-\xi^2} - \mathrm{i}\omega \frac{\xi}{\sqrt{1-\xi^2}} \frac{\partial \xi}{\partial \omega}. \tag{6.68}$$

**Damping Partial Derivative**

The partial derivatives of the damping terms depend on the selection of the Rayleigh damping coefficients $\alpha_R$ and $\beta_R$. A constant choice for $\alpha_R$ and $\beta_R$ leads to

$$\frac{\partial \mathbf{C}}{\partial \omega} = \mathbf{0}$$
$$\frac{\partial \delta}{\partial \omega} = 2\beta_R \omega$$
$$\frac{\partial \xi}{\partial \omega} = \frac{\beta_R \omega^2 - \alpha_R}{2\omega^2}.$$

Otherwise, if a constant Q-factor is selected, the derivatives of the damping terms are:

$$\frac{\partial \mathbf{C}}{\partial \omega} = \frac{\partial \alpha_R}{\partial \omega}\mathbf{M} + \frac{\partial \beta_R}{\partial \omega}\mathbf{K}$$
$$\frac{\partial \delta}{\partial \omega} = \frac{1}{Q}$$
$$\frac{\partial \xi}{\partial \omega} = 0.$$

There are two possibilities to achieve a constant Q-factor, and this choice affects the derivative of the damping matrix $\mathbf{C}$:

$$\text{if} \quad \alpha_R = 0 \quad \text{and} \quad \beta_R = \frac{1}{\omega Q} \implies \frac{\partial \mathbf{C}}{\partial \omega} = -\frac{1}{\omega^2 Q}\mathbf{K}$$
$$\text{if} \quad \alpha_R = \frac{\omega}{Q} \quad \text{and} \quad \beta_R = 0 \implies \frac{\partial \mathbf{C}}{\partial \omega} = \frac{1}{Q}\mathbf{M}.$$

The nonlinear force contribution $\boldsymbol{f_m}$ does not depend on the eigenfrequency $\omega$, thus $\frac{\partial \boldsymbol{f_m}}{\partial \omega} = \mathbf{0}$.

The partial derivatives of vectors $\boldsymbol{V_m}$ and $\dot{\boldsymbol{V}}_{\boldsymbol{m}}$ are

$$\frac{\partial \boldsymbol{V_m}}{\partial \omega} = \sum_{j=1}^{2} \sum_{\substack{\boldsymbol{u},\boldsymbol{k} \in \mathbb{N}^2 \\ 1<u,k<m \\ \boldsymbol{u}+\boldsymbol{k}-\boldsymbol{e}_j=\boldsymbol{m}}} u_j \boldsymbol{w_u} \frac{\partial R_{\boldsymbol{k}}^j}{\partial \omega} \tag{6.69}$$

$$\frac{\partial \dot{V}_m}{\partial \omega} = \sum_{j=1}^{2} \sum_{\substack{u,k \in \mathbb{N}^2 \\ 1<u,k<m \\ u+k-e_j=m}} u_j \left( \frac{\partial \dot{w}_u}{\partial \omega} R_k^j + \dot{w}_u \frac{\partial R_k^j}{\partial \omega} \right). \tag{6.70}$$

The partial derivatives of vector $C_m$ is

$$\begin{aligned} \frac{\partial C_m}{\partial \omega} &= -\mathbf{M}\frac{\partial \dot{V}_m}{\partial \omega} - \left( \frac{\partial \mathbf{\Lambda}_m}{\partial \omega}\mathbf{M} + \frac{\partial \mathbf{C}}{\partial \omega} \right) V_m \\ &\quad - (\Lambda_m \mathbf{M} + \mathbf{C}) \frac{\partial V_m}{\partial \omega}. \end{aligned} \tag{6.71}$$

The partial derivative of the reduced dynamics coefficient is

$$\frac{\partial R_m^j}{\partial \omega} = \frac{\varphi^T \frac{\partial C_m}{\partial \omega}}{\Lambda_m + \lambda_j + \delta} - R_m^j \frac{\frac{\partial \Lambda_m}{\partial \omega} + \frac{\partial \lambda_j}{\partial \omega} + \frac{\partial \delta}{\partial \omega}}{\Lambda_m + \lambda_j + \delta}. \tag{6.72}$$

The partial derivative of the right-hand side of the cohomological equation is

$$\frac{\partial h_m}{\partial \omega} = \frac{\partial C_m}{\partial \omega} + \sum_{j=1}^{2} \left( \frac{\partial D_m^j}{\partial \omega} R_m^j + D_m^j \frac{\partial R_m^j}{\partial \omega} \right). \tag{6.73}$$

where

$$\frac{\partial D_m^j}{\partial \omega} = -\left[ \left( \frac{\partial \Lambda_m}{\partial \omega} + \frac{\partial \lambda_j}{\partial \omega} \right) \mathbf{M} + \frac{\partial \mathbf{C}}{\partial \omega} \right] \varphi, \quad j = 1, 2. \tag{6.74}$$

The partial derivative of the left-hand side of the cohomological equation is

$$\frac{\partial \mathbf{L}_m}{\partial \omega} = (\mathbf{C} + 2\Lambda_m \mathbf{M}) \frac{\partial \Lambda_m}{\partial \omega} + \Lambda_m \frac{\partial \mathbf{C}}{\partial \omega}. \tag{6.75}$$

The partial derivative of the manifold velocity coefficient is

$$\frac{\partial \dot{w}_m}{\partial \omega} = \frac{\partial \Lambda_m}{\partial \omega} w_m + \varphi \sum_{j=1}^{2} \frac{\partial R_m^j}{\partial \omega} + \frac{\partial V_m}{\partial \omega}. \tag{6.76}$$

The partial derivative of the non-autonomous SSM coefficient $S_0$ is

$$\frac{\partial S_0}{\partial \omega} = -S_0 \frac{\frac{\partial \Lambda_0}{\partial \omega} + \frac{\partial \lambda}{\partial \omega} + \frac{\partial \delta}{\partial \omega}}{\Lambda_0 + \lambda + \delta}, \tag{6.77}$$

where

$$\frac{\partial \Lambda_0}{\partial \omega} = \mathrm{i}\,\mathrm{Im}\left\{\frac{\partial \lambda}{\partial \omega}\right\}. \tag{6.78}$$

Finally, the partial derivatives of $a$, $\Omega$, and $x$ are

$$\frac{\partial a}{\partial \omega} = \frac{1}{2}\sum_{m\in\mathbb{N}}\left(\frac{\partial \gamma_m^2}{\partial \omega} + \frac{\partial \gamma_m^1}{\partial \omega}\right)\rho^m \tag{6.79}$$

$$\frac{\partial \Omega}{\partial \omega} = \frac{1}{2}\mathrm{i}\sum_{m>1}\left(\frac{\partial \gamma_m^2}{\partial \omega} - \frac{\partial \gamma_m^1}{\partial \omega}\right)\rho^{m-1} \tag{6.80}$$

$$\frac{\partial x}{\partial \omega} = 0. \tag{6.81}$$

### 6.6.5 Partial Derivatives with Respect to the Design Variables

The partial derivative of the damping matrix $\mathbf{C}$ with respect to the design variable $\mu$ is

$$\frac{\partial \mathbf{C}}{\partial \mu} = \alpha_R\frac{\partial \mathbf{M}}{\partial \mu} + \beta_R\frac{\partial \mathbf{K}}{\partial \mu}, \tag{6.82}$$

where the values of $\alpha_R$ and $\beta_R$ do not depend on $\mu$.

Using tensor notation, the partial derivative of the nonlinear force contribution is

$$\frac{\partial f_{\boldsymbol{m}}^i}{\partial \mu} = \sum_{\substack{\boldsymbol{u},\boldsymbol{k}\in\mathbb{N}^2\\ \boldsymbol{u}+\boldsymbol{k}=\boldsymbol{m}}}\frac{\partial T_2^{ijk}}{\partial \mu}w_{\boldsymbol{u}}^j w_{\boldsymbol{k}}^k + \sum_{\substack{\boldsymbol{u},\boldsymbol{k},\boldsymbol{l}\in\mathbb{N}^2\\ \boldsymbol{u}+\boldsymbol{k}+\boldsymbol{l}=\boldsymbol{m}}}\frac{\partial T_3^{ijkl}}{\partial \mu}w_{\boldsymbol{u}}^j w_{\boldsymbol{k}}^k w_{\boldsymbol{l}}^l. \tag{6.83}$$

The partial derivative of vectors $\boldsymbol{V}_{\boldsymbol{m}}$ and $\dot{\boldsymbol{V}}_{\boldsymbol{m}}$ are

$$\frac{\partial \boldsymbol{V}_{\boldsymbol{m}}}{\partial \mu} = \sum_{j=1}^{2}\sum_{\substack{\boldsymbol{u},\boldsymbol{k}\in\mathbb{N}^2\\ 1<u,k<m\\ \boldsymbol{u}+\boldsymbol{k}-\boldsymbol{e}_j=\boldsymbol{m}}}\boldsymbol{w}_{\boldsymbol{u}}u_j\frac{\partial R_{\boldsymbol{k}}^j}{\partial \mu} \tag{6.84}$$

$$\frac{\partial \dot{\boldsymbol{V}}_{\boldsymbol{m}}}{\partial \mu} = \sum_{j=1}^{2}\sum_{\substack{\boldsymbol{u},\boldsymbol{k}\in\mathbb{N}^2\\ 1<u,k<m\\ \boldsymbol{u}+\boldsymbol{k}-\boldsymbol{e}_j=\boldsymbol{m}}}u_j\left[\frac{\partial \dot{\boldsymbol{w}}_{\boldsymbol{u}}}{\partial \mu}R_{\boldsymbol{k}}^j + \dot{\boldsymbol{w}}_{\boldsymbol{u}}\frac{\partial R_{\boldsymbol{k}}^j}{\partial \mu}\right]. \tag{6.85}$$

The partial derivatives of vector $\boldsymbol{C}_{\boldsymbol{m}}$ is

$$\begin{aligned}\frac{\partial \boldsymbol{C}_{\boldsymbol{m}}}{\partial \mu} &= -\frac{\partial \mathbf{M}}{\partial \mu}\dot{\boldsymbol{V}}_{\boldsymbol{m}} - \mathbf{M}\frac{\partial \dot{\boldsymbol{V}}_{\boldsymbol{m}}}{\partial \mu} - (\Lambda_{\boldsymbol{m}}\mathbf{M} + \mathbf{C})\frac{\partial \boldsymbol{V}_{\boldsymbol{m}}}{\partial \mu} \\ &\quad - \left(\Lambda_{\boldsymbol{m}}\frac{\partial \mathbf{M}}{\partial \mu} + \frac{\partial \mathbf{C}}{\partial \mu}\right)\boldsymbol{V}_{\boldsymbol{m}} - \frac{\partial \boldsymbol{f}_{\boldsymbol{m}}}{\partial \mu}.\end{aligned} \tag{6.86}$$

The partial derivative of the reduced dynamics coefficient is

$$\frac{\partial R_{\boldsymbol{m}}^{j}}{\partial \mu} = \frac{\boldsymbol{\varphi}^{T}\frac{\partial \boldsymbol{C}_{\boldsymbol{m}}}{\partial \mu}}{\Lambda_{\boldsymbol{m}} + \lambda_j + \delta}. \tag{6.87}$$

The partial derivative of the right-hand side of the cohomological equation is

$$\frac{\partial \boldsymbol{h}_{\boldsymbol{m}}}{\partial \mu} = \frac{\partial \boldsymbol{C}_{\boldsymbol{m}}}{\partial \mu} + \sum_{j=1}^{2}\left(\frac{\partial \boldsymbol{D}_{\boldsymbol{m}}^{j}}{\partial \mu}R_{\boldsymbol{m}}^{j} + \boldsymbol{D}_{\boldsymbol{m}}^{j}\frac{\partial R_{\boldsymbol{m}}^{j}}{\partial \mu}\right), \tag{6.88}$$

where

$$\frac{\partial \boldsymbol{D}_{\boldsymbol{m}}^{j}}{\partial \mu} = -\left[\left(\Lambda_{\boldsymbol{m}} + \lambda_j\right)\frac{\partial \mathbf{M}}{\partial \mu} + \frac{\partial \mathbf{C}}{\partial \mu}\right]\boldsymbol{\varphi}. \tag{6.89}$$

The partial derivative of the left-hand side of the cohomological equation is

$$\frac{\partial \mathbf{L}_{\boldsymbol{m}}}{\partial \mu} = \frac{\partial \mathbf{K}}{\partial \mu} + \Lambda_{\boldsymbol{m}}\frac{\partial \mathbf{C}}{\partial \mu} + \Lambda_{\boldsymbol{m}}^{2}\frac{\partial \mathbf{M}}{\partial \mu}. \tag{6.90}$$

The partial derivative of the manifold velocity coefficient is

$$\frac{\partial \dot{\boldsymbol{w}}_{\boldsymbol{m}}}{\partial \mu} = \sum_{j=1}^{2}\frac{\partial R_{\boldsymbol{m}}^{j}}{\partial \mu}\boldsymbol{\varphi} + \frac{\partial \boldsymbol{V}_{\boldsymbol{m}}}{\partial \mu}. \tag{6.91}$$

Finally, the partial derivatives of $a$, $\Omega$, and $x$ are

$$\frac{\partial a}{\partial \mu} = \frac{1}{2}\sum_{m\in\mathbb{N}}\left(\frac{\partial \gamma_m^2}{\partial \mu} + \frac{\partial \gamma_m^1}{\partial \mu}\right)\rho^{m} \tag{6.92}$$

$$\frac{\partial \Omega}{\partial \mu} = \frac{1}{2}\mathrm{i}\sum_{m>1}\left(\frac{\partial \gamma_m^2}{\partial \mu} - \frac{\partial \gamma_m^1}{\partial \mu}\right)\rho^{m-1} \tag{6.93}$$

$$\frac{\partial x}{\partial \mu} = 0. \tag{6.94}$$

## References

1. Christensen PW, Klarbring A (2008) An introduction to structural optimization, vol 153, 1st edn. Springer, Dordrecht. https://doi.org/10.1007/978-1-4020-8666-3
2. Manzoni A, Quarteroni A, Salsa S (2021) Optimal control of partial differential equations, vol. 207. Springer, Berlin. https://doi.org/10.1007/978-3-030-77226-0
3. Pozzi M, Marconi J, Jain S, Li M, Braghin F (2025) Adjoint sensitivities for the optimization of nonlinear structural dynamics via spectral submanifolds. Proc R Soc A Math Phys Eng Sci 481:20250244. https://doi.org/10.1098/rspa.2025.0244
4. Wirtinger W (1927) Zur formalen theorie der funktionen von mehr komplexen veränderlichen. Math Ann 97:357–375. https://doi.org/10.1007/BF01447872

# Chapter 7
# Results and Discussion

**Abstract** This chapter presents numerical examples demonstrating the SSM-based optimization framework. Case studies of parametric optimization, including a MEMS gyroscope, showcase the method's applicability. Additional topology optimization examples highlight the framework's efficiency and scalability.

## 7.1 Introduction

This chapter presents the application of the Spectral Submanifold-based optimization framework introduced in Chap. 5, highlighting its efficiency and versatility in both parametric and topology optimization problems. All case studies are implemented in MATLAB using `YetAnotherFEcode` [5], with the sensitivity analysis carried out via the adjoint method described in Chap. 6.

In all examples, the material is assumed to be polysilicon [1], characterized by Young's modulus $E = 161\,\text{GPa}$, Poisson's ratio $\nu = 0.23$, and density $\rho = 2330\,\text{kg/m}^3$. A plane stress assumption is adopted, with a uniform out-of-plane thickness of 24 $\mu$m. To ensure correct identification of the target modes during optimization, the Modal Assurance Criterion (MAC) [7] is applied as described in Sect. 4.3.1.

Unless otherwise specified, all computational times reported in this chapter were measured on a Windows workstation equipped with an Intel Core i9-13900KS CPU @ 3.20 GHz and 128 GB of RAM @ 4200 MT/s.

## 7.2 Parametric Optimization

This section focuses on parametric optimization, illustrating the approach through two representative case studies: a von Kármán beam and a single-axis MEMS gyroscope, previously introduced in Chap. 3. The optimization problems are solved using MATLAB's gradient-based optimizer `fmincon`, configured with the `interior-point` algorithm.



M. Pozzi, F. Braghin, *Structural Optimization for Nonlinear Dynamics*,
PoliMI SpringerBriefs, https://doi.org/10.1007/978-3-032-22600-6_7

### 7.2.1 *Optimization of a von Kármán Beam*

In this first example, we present the case of a geometrically nonlinear, undamped, clamp-clamp von Kármán beam. The structure is described by a finite element model consisting of 10 von Kármán elements (27 degrees of freedom).

As design variables, we take the beam length $L$ and thickness $h$. Additionally, to enable shape optimization of the beam, we allow its shape to deviate from a straight path by superimposing a finite set of sinusoidal modes. For a beam aligned with the $x$-axis, the perturbed vertical coordinate is expressed as:

$$y(x) = y_0(x) + \sum_{k=1}^{N_A} A_k \sin\left(\frac{k\pi}{L}x\right), \tag{7.1}$$

where $y_0$ is the original vertical coordinate, $L$ is the current beam length, and $A_k$ are the harmonic amplitudes. The sinusoidal modes are chosen to preserve the position of the nodes at the beam ends, i.e., where the boundary conditions are applied. In this example, two harmonics are used ($N_A = 2$), and thus the design is governed by four parameters:

$$\boldsymbol{\mu} = \{L,\ h,\ A_1,\ A_2\},$$

where the initial values and the bounds are listed in Table 7.1.

To obtain a softening response for the first mode, we define two target points for the backbone curve as

$$(\Omega_1,\ x_1) = (1.0\,\omega_0,\ 0.2\,h_0)$$
$$(\Omega_2,\ x_2) = (0.9\,\omega_0,\ 0.4\,h_0),$$

where $\omega_0 = 85.4$ kHz is the first eigenfrequency of the nominal beam and $h_0$ its nominal thickness. The optimization problem is then formulated as

$$\begin{aligned} \min_{\boldsymbol{\mu}} \quad & A_1 + A_2 \\ \text{s.t.} \quad & \Omega(x_1) - \Omega_1 = 0 \\ & \Omega(x_2) - \Omega_2 = 0 \\ & \boldsymbol{\mu}_\mathrm{l} \leq \boldsymbol{\mu} \leq \boldsymbol{\mu}_\mathrm{u}, \end{aligned} \tag{7.2}$$

where the objective function is the sum of the amplitudes of the two harmonics.

**Table 7.1** Parameters used in the von Kármán beam optimization problem (7.2)

| | $L$ | $h$ | $A_1$ | $A_2$ |
|---|---|---|---|---|
| Initial values [$\mu$m] | 1000 | 10.0 | 0.00 | 0.00 |
| Lower bounds [$\mu$m] | 500 | 1.00 | 0.00 | 0.00 |
| Upper bounds [$\mu$m] | 1500 | 50.0 | 1.00 | 1.00 |
| Optimal values [$\mu$m] | 1000 | 6.38 | 7.90 | 0.00 |

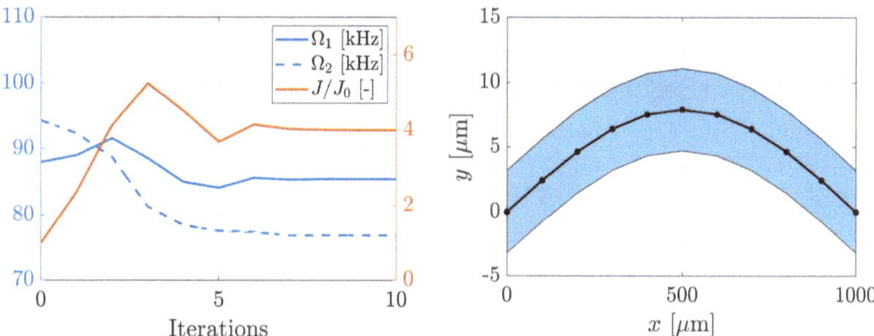


**Fig. 7.1** Optimization history (left) and optimal shape (right) of the von Kármán beam optimization problem (7.2)

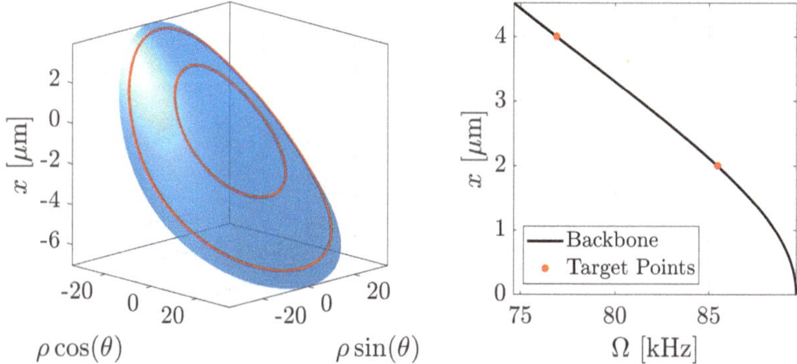


**Fig. 7.2** Spectral submanifold (left) and conservative backbone curve (right) of the von Kármán beam in Fig. 7.1

Problem (7.2) is solved using the routine described in Sect. 5.2. The optimization reaches convergence after 10 iterations in 5 seconds, with an average time of 0.5 seconds per iteration. At iteration 4, the SSM expansion order is increased from 3 to 5 to keep the residual of the invariance equation below the threshold $\epsilon_{\max} = 0.1$.

Figures 7.1 and 7.2 presents the complete optimization results, including the optimization history, the final layout, and the corresponding invariant manifold and backbone curve.

### 7.2.2 *MEMS Gyroscope with Drive Linearization*

At the end of Chap. 3, we emphasized the importance of accounting for nonlinear effects in the design process, using a single-axis MEMS gyroscope as a case study. The device was optimized to match target drive and sense frequencies of 28 kHz and 26 kHz, respectively, while maximizing the first spurious frequency. Subsequently, the nonlinear dynamic response of the drive mode was analyzed using the $\mathcal{O}(3)$ SSM reduction, revealing a hardening-type nonlinearity. Under nominal forcing, this nonlinearity caused the peak response to shift to approximately 32.2 kHz, with an oscillation amplitude of 2.22 μm. As a result, the actual frequency mismatch between drive and sense frequencies increased to 6.2 kHz, a significantly higher value than the designed one of 2 kHz. This variation affects the overall performance of the device, potentially reducing both its sensitivity and bandwidth [6].

To address this issue, we aim to obtain a drive response that is as linear as possible. This result can be achieved by minimizing the magnitude of the third-order backbone coefficient $\text{Im}\{\gamma_3\}$. To this end, the optimization problem is formulated as:

$$\begin{aligned} \min_{\boldsymbol{\mu}} \quad & \text{Im}\{\gamma_3\}^2 \\ \text{s.t.} \quad & \omega_d - 28\,\text{kHz} = 0 \\ & \omega_s - 26\,\text{kHz} = 0 \\ & \boldsymbol{\mu}_\text{l} \leq \boldsymbol{\mu} \leq \boldsymbol{\mu}_\text{u}, \end{aligned} \tag{7.3}$$

where $\boldsymbol{\mu}$ denotes the vector of design parameters.

As in Sect. 3.2, the Multi-Point Constraint (MPC) approach [3] is adopted to model the MEMS gyroscope as an assembly of rigid bodies, the drive and sense frames, connected by elastic elements; these elastic connections are modeled using von Kármán beam elements to capture geometric nonlinearities. The finite element model has a total of 366 degrees of freedom. The device geometry is parameterized using the length and width of the drive beams ($L_d$ and $W_d$), the length and width of the folded sense beams ($L_s$ and $W_s$), and the width of the connecting segments between the folded sense beams ($W_c$).

To enable shape optimization of the drive beams, their shapes are defined by superimposing a finite set of sinusoidal modes, as described in Eq. (7.1). Independent sets of harmonic coefficients are assigned to the two top ($A_k^t$) and two bottom ($A_k^b$) drive beams, with $k = 1, \ldots, 5$. Altogether, the design is governed by 15 parameters:

$$\boldsymbol{\mu} = \{L_d,\ L_s,\ W_d,\ W_s,\ W_c,\ A_1^t,\ \ldots,\ A_5^t,\ A_1^b,\ \ldots,\ A_5^b\}. \tag{7.4}$$

Table 7.2 gives the initial values and the bounds for the parameters, expressed as deviations with respect to the nominal values of the initial layout (Table 3.1). In this example, the amplitudes of the harmonics of the top drive beams are imposed to be equal to those of the bottom drive beams:

$$A_k = A_k^t = A_k^b \quad \forall k = 1, \ldots, 5.$$

Using a fixed $\mathcal{O}(3)$ SSM reduction, the optimization problem (7.3) converges after 65 iterations in 3 minutes and 45 seconds, with an average iteration time of 3.5 seconds. The optimal parameters are reported in Table 7.3.

**Table 7.2** Initial parameters and absolute bounds used in the MEMS optimization problem (7.3). All the values are expressed as deviations (e.g., $L_d = L_{d,0}+\Delta L_d$) from their nominal values of the initial layout (Table 3.1). All the amplitudes of the drive beam harmonics share the same nominal value of 0.00 µm, initial value of 0.00 µm, and absolute bound of 3.00 µm

| | $\Delta L_d$ | $\Delta L_s$ | $\Delta W_d$ | $\Delta W_s$ | $\Delta W_c$ |
|---|---|---|---|---|---|
| Initial values [$\mu$m] | 0.00 | 0.00 | 0.00 | 0.00 | 0.00 |
| Absolute bounds [$\mu$m] | 25.0 | 25.0 | 3.00 | 3.00 | 3.00 |

**Table 7.3** Optimal results of the MEMS gyroscope optimization problem (7.3)

| $\Delta L_d$ [$\mu$m] | $\Delta L_s$ [$\mu$m] | $\Delta W_d$ [$\mu$m] | $\Delta W_s$ [$\mu$m] | $\Delta W_c$ [$\mu$m] |
|---|---|---|---|---|
| 25.0 | −25.0 | 0.80 | −0.97 | 0.98 |
| $\Delta A_1$ [$\mu$m] | $\Delta A_2$ [$\mu$m] | $\Delta A_3$ [$\mu$m] | $\Delta A_4$ [$\mu$m] | $\Delta A_5$ [$\mu$m] |
| −3.00 | −3.00 | 3.00 | 3.00 | 3.00 |

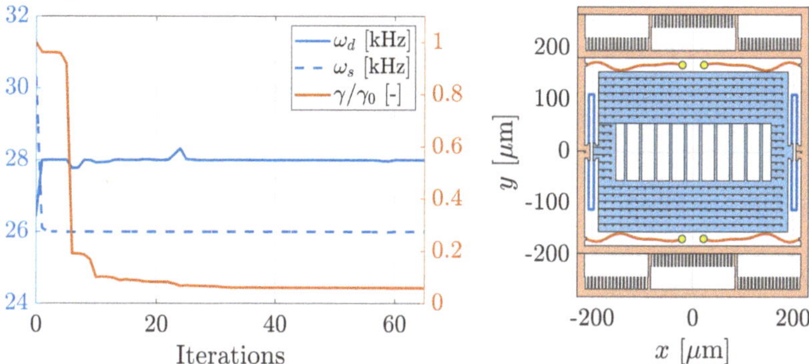


**Fig. 7.3** Optimization history (left) and optimal layout (right) of the MEMS gyroscope optimization problem (7.3)

The optimization history and the resulting optimal layout are shown in Fig. 7.3, along with the corresponding frequency response (Fig. 7.4). For comparison, the response of this new layout (in red) is plotted alongside that of the reference layout (in blue), which was obtained by solving problem (3.1) in Sect. 3.3, i.e., without accounting for geometric nonlinearities.

At a displacement amplitude of 2.22 µm, the reference layout yields a drive frequency of 32.2 kHz, which overshoots the target value of 28 kHz by approximately 15%. In contrast, the new optimized layout achieves a drive frequency of 28.3 kHz at the same amplitude, deviating from the target by only 1.1%. This more linear response comes at the cost of increased stiffness: the new layout requires a higher actuation force of 0.36 µN to reach the same displacement amplitude, compared to 0.30 µN required by the reference design (Sect. 3.3.1).

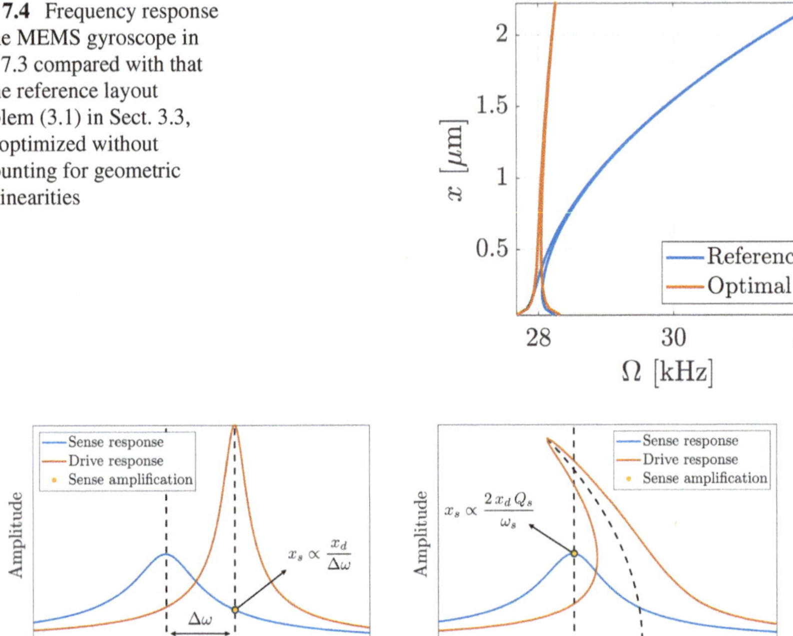


**Fig. 7.4** Frequency response of the MEMS gyroscope in Fig. 7.3 compared with that of the reference layout problem (3.1) in Sect. 3.3, i.e., optimized without accounting for geometric nonlinearities

**Fig. 7.5** Comparison between the typical linear frequency response of a MEMS gyroscope with a prescribed frequency mismatch (left) and the proposed nonlinear strategy enabling mode matching (right)

## 7.2.3 *MEMS Gyroscope with Mode-Matching*

In Sect. 7.2.2, we used the SSM-based optimization to suppress the nonlinearities in the drive mode of a MEMS gyroscope. In this section, we take a different approach and aim to exploit these nonlinearities instead. As discussed in Sect. 1.1, a prescribed frequency mismatch $\Delta\omega$ between the sense and drive modes is typically introduced to enhance the robustness of MEMS gyroscopes. However, this comes at the expense of reduced sense mode amplification, as illustrated in Fig. 7.5. Here, we investigate the potential of utilizing nonlinear effects to achieve mode matching while preserving the prescribed frequency mismatch, thereby enhancing the device's sensitivity while maintaining its robustness. A conceptual illustration of this idea is shown in Fig. 7.5.

To exploit nonlinearities for mode matching, we formulate the following optimization problem:

$$
\begin{aligned}
\max_{\boldsymbol{\mu}} \quad & \omega_{sp} \\
\text{s.t.} \quad & \omega_d - 28\,\text{kHz} = 0 \\
& \omega_s - 26\,\text{kHz} = 0 \\
& \Omega(\varepsilon_{\text{nominal}}) - 26\,\text{kHz} = 0 \\
& 2\,\mu\text{m} - x(\varepsilon_{\text{nominal}}) \leq 0 \\
& \boldsymbol{\mu}_\text{l} \leq \boldsymbol{\mu} \leq \boldsymbol{\mu}_\text{u},
\end{aligned}
\tag{7.5}
$$

where the objective function maximizes the first spurious frequency $\omega_{sp}$. The constraints are used to enforce a prescribed frequency mismatch between the drive $\omega_d = 28\,\text{kHz}$ and the sense $\omega_s = 26\,\text{kHz}$ modes, while ensuring nonlinear frequency matching at the nominal forcing amplitude $\varepsilon_{\text{nominal}} = 0.30\,\mu\text{N}$ defined in Sect. 3.3.1. Additionally, the response amplitude at $\varepsilon_{\text{nominal}}$ must be greater than $2\,\mu\text{m}$ to ensure a reasonable drive displacement.

Using the same problem settings described in Sect. 7.2.2, the optimization problem (7.5) converges in 12 minutes and 25 seconds after 45 iterations, yielding the optimal parameters reported in Table 7.4. The average iteration time is 17 seconds. The optimization history and the optimal layout are shown in Figs. 7.6 and 7.7, along with the evolution of the residual error and the nonlinear frequency response of the drive mode in Fig. 7.8. The response peak occurs at the same frequency as the sense mode, thereby achieving mode matching.

The optimization begins with a $\mathcal{O}(5)$ SSM reduction, and the expansion order is kept constant throughout the optimization, as the residual error remains within acceptable bounds during the entire process (Fig. 7.8). For completeness, we tried to start the optimization with an $\mathcal{O}(3)$ SSM reduction. However, in this case, the residual error exceeded the upper threshold already at the first iteration, indicating that both the frequency response and its sensitivities were unreliable. As a result, the optimization converged prematurely to an infeasible point after just one iteration. This outcome highlights the importance of selecting an appropriate expansion order and error tolerance, ideally based on a preliminary assessment of the system's behavior.

**Table 7.4** Optimal results of the MEMS gyroscope optimization problem (7.5)

| $\Delta L_d$ [$\mu$m] | $\Delta L_s$ [$\mu$m] | $\Delta W_d$ [$\mu$m] | $\Delta W_s$ [$\mu$m] | $\Delta W_c$ [$\mu$m] |
|---|---|---|---|---|
| −22.1 | 17.1 | −1.00 | 0.11 | −1.00 |
| $\Delta A_1^t$ [$\mu$m] | $\Delta A_2^t$ [$\mu$m] | $\Delta A_3^t$ [$\mu$m] | $\Delta A_4^t$ [$\mu$m] | $\Delta A_5^t$ [$\mu$m] |
| 0.05 | 1.23 | 0.02 | 0.62 | 0.01 |
| $\Delta A_1^b$ [$\mu$m] | $\Delta A_2^b$ [$\mu$m] | $\Delta A_3^b$ [$\mu$m] | $\Delta A_4^b$ [$\mu$m] | $\Delta A_5^b$ [$\mu$m] |
| −0.05 | 1.26 | −0.02 | 0.64 | −0.01 |

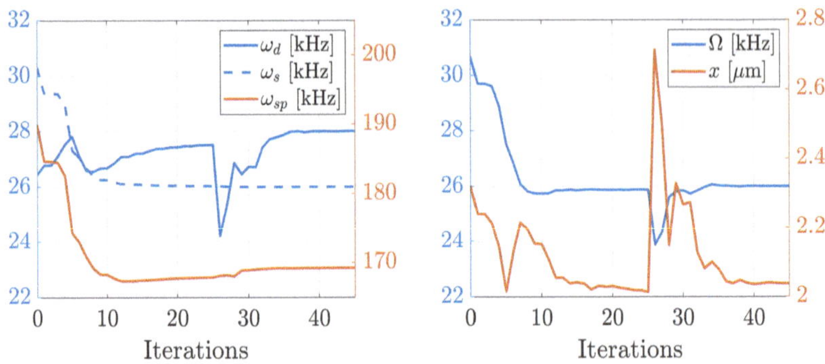


**Fig. 7.6** Optimization history of the MEMS gyroscope optimization problem (7.5)

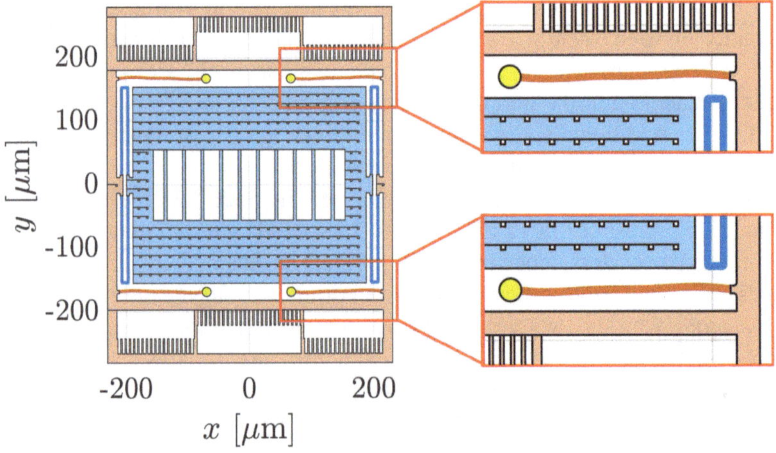


**Fig. 7.7** Optimal layout of the MEMS gyroscope optimization problem (7.5)

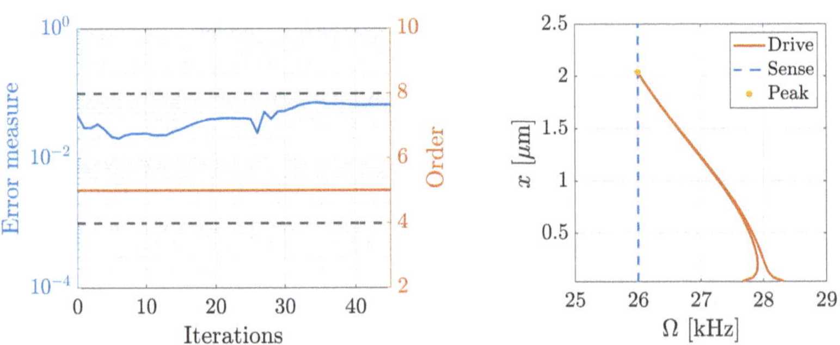


**Fig. 7.8** Evolution of the residual error (left) and nonlinear frequency response of the drive mode (right) of the MEMS gyroscope optimization problem (7.5)

## 7.3 Topology Optimization

This section addresses topology optimization, employing the density-based approach described in Chap. 4 to investigate two nonlinear dynamic design problems. Specifically, the focus is on maximizing the hardening/softening behavior of a single-mass MEMS resonator and tailoring the backbone curve of an Messerschmitt-Bölkow-Blohm (MBB) beam.

### 7.3.1 *Hardening/Softening Optimization of a MEMS Resonator*

The first topology optimization example is a single-mass MEMS resonator designed for high-frequency applications. The design domain ($1000 \times 500\,\mu\text{m}$) is discretized using a $200 \times 100$ grid. Problem settings and boundary conditions are described in Fig. 7.9. The black region represents the proof mass that remains constant for the entire optimization. Fixed boundary conditions are applied to both the left and right sides of the domain.

We define $\omega_y$ as the frequency corresponding to the mode shape with the maximum displacement of the proof mass along the $y$ direction. Additionally, we define $\omega_x$ as the frequency corresponding to the mode shape with the oscillations of the proof mass along the $x$ direction.

The hardening/softening behavior of the resonator is described using the third-order backbone coefficient $\text{Im}\{\gamma_3\}$, which can be computed using a $\mathcal{O}(3)$ SSM reduction (Sect. 5.2.1). A positive value of $\text{Im}\{\gamma_3\}$ indicates a hardening behavior, while a negative value indicates softening. Ideally, a zero value of $\text{Im}\{\gamma_3\}$ means a perfectly linear behavior, at least at low amplitudes.

To design the resonator, we aim to impose the eigenfrequency $\omega_y$ to the target value of 1000 kHz with a tolerance of 0.1%. Furthermore, the eigenfrequency $\omega_x$ is used as a connectivity constraint [2, 4] to avoid gray transition regions and solid islands of material. Moreover, a maximum volume of 60% of the available one is added to limit the total weight of the device.

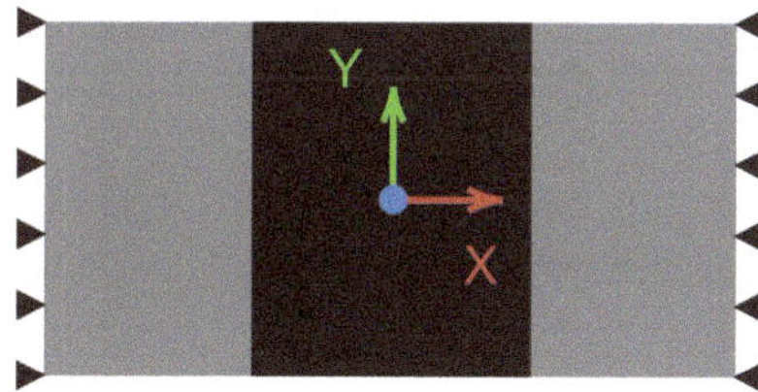


**Fig. 7.9** Problem settings and initial conditions for the MEMS resonator

The first example aims at maximizing the hardening behavior in the resonator response. To do so, the following optimization problem is defined to maximize the value of $\mathrm{Im}\{\gamma_3\}$:

$$\begin{aligned} \max_{\boldsymbol{\mu}} \quad & \mathrm{Im}\{\gamma_3\} \\ \text{s.t.} \quad & (\omega_y/\omega_{y,\text{target}} - 1)^2 - \epsilon^2 \leq 0 \\ & \omega_{x,\min} - \omega_x \leq 0 \\ & V - 60\% \leq 0 \\ & \mathbf{0} \leq \boldsymbol{\mu} \leq \mathbf{1}, \end{aligned} \tag{7.6}$$

where the lower bound for the $\omega_x$ frequency is selected as $\omega_{x,\min} = 2000\,\text{kHz}$. The value of $\epsilon$ represents the relative error of $\omega_y$ with respect to $\omega_{y,\text{target}}$. For this problem, $\epsilon = 10^{-3}$ is used, meaning that the constraint is satisfied if the value of $\omega_y$ is within 0.1% of the target value (between 999 and 1001 kHz).

The optimal layout and the corresponding backbone curve are shown in Fig. 7.10. The optimization converges to a design with $\omega_y = 999.0\,\text{kHz}$, $\omega_x = 2000\,\text{kHz}$, and $\mathrm{Im}\{\gamma_3\} = 1.45 \cdot 10^{-4}$, where the positive value of the third-order backbone coefficient indicates the desired hardening behavior. The process completes in 57 minutes and 10 seconds, requiring 128 iterations with an average of 26.8 seconds per iteration.

The second optimization problem is defined to maximize the softening behavior by minimizing the value of $\mathrm{Im}\{\gamma_3\}$ to obtain a negative value. Similarly to (7.6), the optimization problem is formulated as:

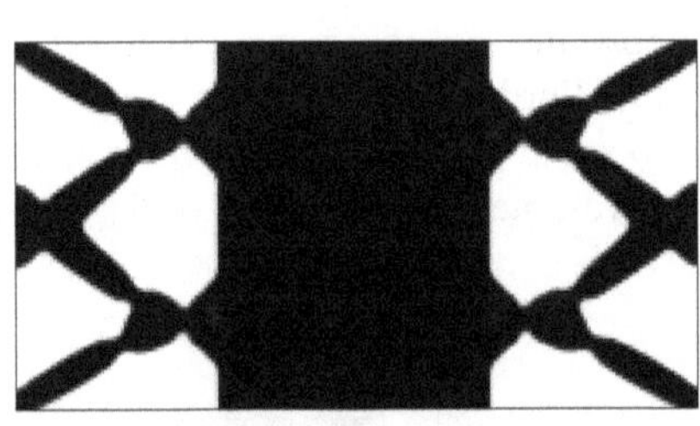

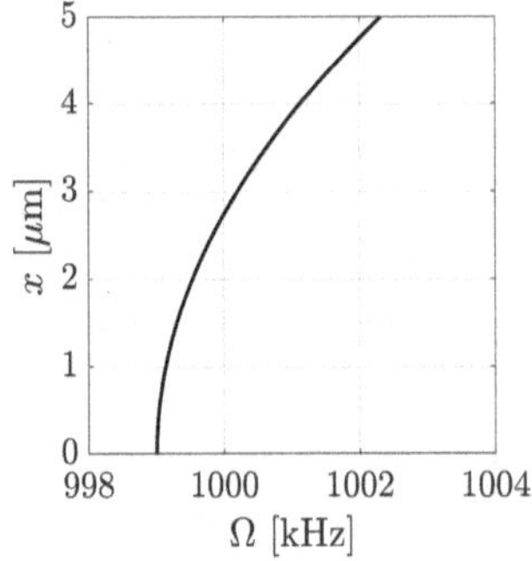


**Fig. 7.10** Optimal layout (left) and corresponding backbone curve (right) of the MEMS resonator hardening maximization (7.6)

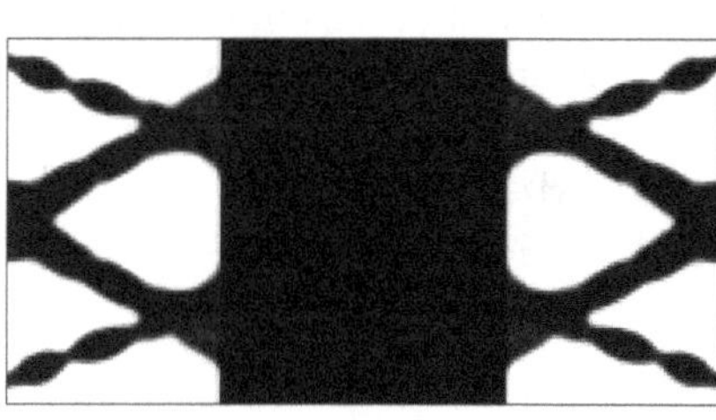

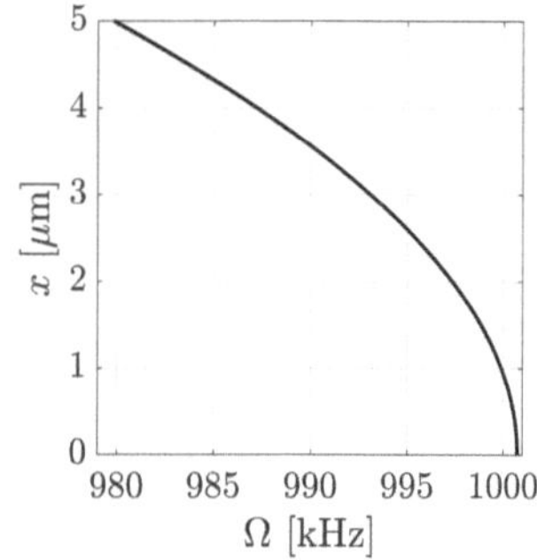


**Fig. 7.11** Optimal layout (left) and corresponding backbone curve (right) of the MEMS resonator softening maximization (7.7).

**Fig. 7.12** Problem settings and initial conditions for the MBB beam.

$$
\begin{aligned}
\min_{\mu} \quad & \mathrm{Im}\{\gamma_3\} \\
\text{s.t.} \quad & (\omega_y/\omega_{y,\text{target}} - 1)^2 - \epsilon^2 \leq 0 \\
& \omega_{x,\min} - \omega_x \leq 0 \\
& V - 60\% \leq 0 \\
& \mathbf{0} \leq \boldsymbol{\mu} \leq \mathbf{1},
\end{aligned}
\tag{7.7}
$$

where all the parameters are the same as in problem (7.6).

The optimal layout and the corresponding backbone curve are shown in Fig. 7.11. The optimization converges to a design with $\omega_y = 1000.7$ kHz, $\omega_x = 2000$ kHz, and $\mathrm{Im}\{\gamma_3\} = -9.18 \cdot 10^{-4}$, where the negative value of the third-order backbone coefficient indicates the desired softening behavior. The process completes after 116 iterations in 52 minutes and 38 seconds. The average time per iteration is 27.2 seconds, which is aligned with the iteration time of the hardening maximization (7.6).

### 7.3.2 *Backbone Optimization of an MBB Beam*

In the final topology optimization example, we consider the MBB beam configuration shown in Fig. 7.12. The design domain ($500 \times 100\,\mu$m) is discretized using a $200 \times 40$ grid. The black region indicates the proof mass, which remains fixed throughout the optimization process. Fixed boundary conditions are imposed on the left edge of the domain, while only horizontal displacements are constrained on the right edge.

The objective of this case study is to maximize the eigenfrequency $\omega_y$ associated with the mode shape with the maximum displacement of the proof mass along the $y$ direction. In addition, we prescribe a target point on the backbone curve as

$$(\Omega_{\text{target}}, x_{\text{target}}) = (620\,\text{kHz},\ 20\,\mu\text{m}).$$

The corresponding optimization problem is defined as:

$$\begin{aligned} \max_{\mu} \quad & \omega_y \\ \text{s.t.} \quad & (\Omega(x_{\text{target}})/\Omega_{\text{target}} - 1)^2 - \epsilon^2 \leq 0 \\ & V - 50\% \leq 0 \\ & \mathbf{0} \leq \boldsymbol{\mu} \leq \mathbf{1}, \end{aligned} \tag{7.8}$$

where $\epsilon = 10^{-3}$ constraints the relative error of the computed frequency $\Omega(x_{\text{target}})$ to within 0.1% of the target value.

Using a constant-order $\mathcal{O}(3)$ SSM reduction, the optimization converges after 143 iterations, requiring a total of 6 hours, 34 minutes, and 49 seconds, with an average of 2 minutes and 45 seconds per iteration. The optimal layout and the corresponding backbone curve are presented in Fig. 7.13. The solution is characterized by frequencies $\omega_y = 695.0\,\text{kHz}$ and $\Omega(x_{\text{target}}) = 620.0\,\text{kHz}$.

During the optimization, some iterations exceeded 240 seconds, while the others had an average iteration time of approximately 130 seconds (Fig. 7.14). A similar trend was also observed in the other topology optimization examples presented in this section. This discrepancy may be attributed to localized phenomena, such as large displacements within void regions, which generate significant nonlinear internal forces. These forces can lead to numerical instabilities, which in turn slow down the SSM computation steps. A more thorough investigation is needed to fully understand the underlying causes. In this case, addressing this issue could

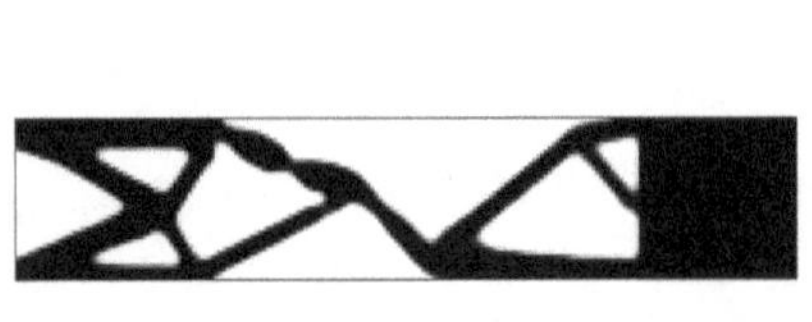

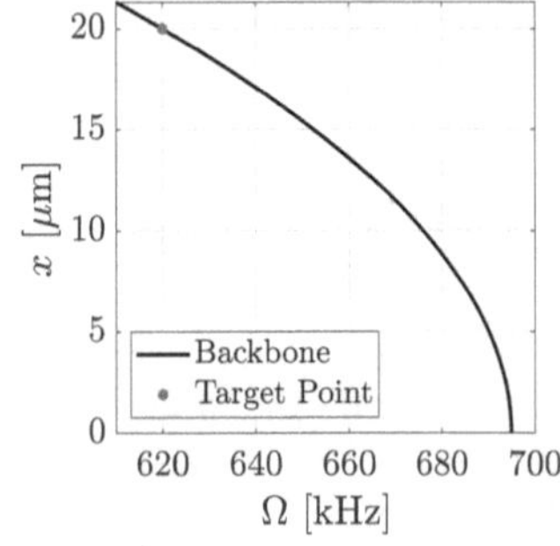


**Fig. 7.13** Optimal layout (left) and corresponding backbone curve (right) of the MBB beam optimization problem (7.8).

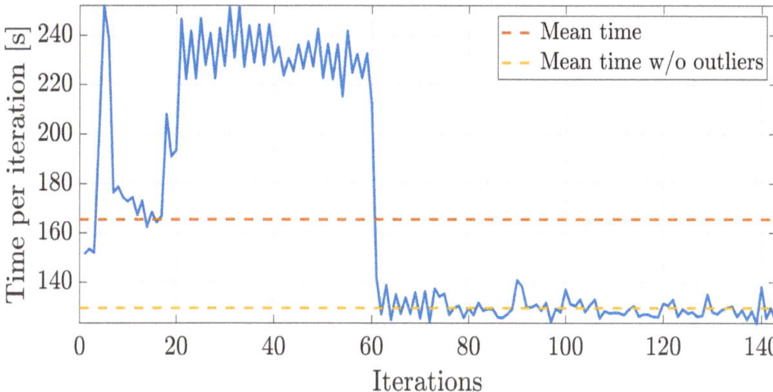


**Fig. 7.14** Time per iterations of the MBB beam optimization problem (7.8). The red line indicates the overall average time. In contrast, the yellow line represents the average time calculated by excluding the times associated with iterations that took more than 150 seconds to complete

potentially reduce the average iteration time by around 20%, thereby decreasing the overall optimization duration. In addition to improving computational efficiency, mitigating these numerical instabilities may also result in a smoother optimization history with fewer iterations.

## References


1. Acar C, Shkel A (2009) MEMS vibratory gyroscopes. Springe, Berlin. https://doi.org/10.1007/978-0-387-09536-3
2. Bonaccorsi G, Pozzi M, Hyun J, Kim HA, Braghin F (2025) Connectivity constraints for eigenvalue reduction in level-set topology optimization. Comput Struct 316:107865. https://doi.org/10.1016/j.compstruc.2025.107865
3. Cook RD, Malkus DS, Plesha ME, Witt RJ (2001) Concepts and applications of finite element analysis, 4th edn. Wiley, Hoboken
4. Cool V, Aage N, Sigmund O (2025) A practical review on promoting connectivity in topology optimization. Struct Multidiscip Optim 68:73. https://doi.org/10.1007/s00158-025-04004-z
5. Jain S, Marconi J, Pozzi M, Tiso P (2026) YetAnotherFEcode (v1.5.0). Zenodo. https://doi.org/10.5281/zenodo.15257966
6. Kempe V (2011) Inertial MEMS. Cambridge University Press, Cambridge. https://doi.org/10.1017/CBO9780511933899
7. Kim TS, Kim YY (2000) Mac-based mode-tracking in structural topology optimization. Comput Struct 74:375–383. https://doi.org/10.1016/S0045-7949(99)00056-5

# Chapter 8
# Conclusion

**Abstract** This chapter summarizes the main findings and contributions of the book, highlighting the potential of the SSM-based framework for optimizing nonlinear mechanical systems. The discussion reflects on the advantages, limitations, and possible extensions of the methodology, and outlines future perspectives on integrating nonlinear dynamics and optimization in engineering design.

The central theme of this work has been the integration of Spectral Submanifold (SSM)-based model reduction with parametric and topology optimization techniques to tailor the nonlinear dynamic response of mechanical systems. By leveraging the analytical structure of SSMs, the approach enables the derivation of closed-form expressions for the nonlinear frequency-amplitude relation associated with a master mode. One of the key strengths of this framework is that it avoids the dependence on user-defined parameters typical of continuation-based methods, such as collocation or Harmonic Balance. Instead, the only parameter required, the expansion order, can be automatically adjusted to ensure the desired accuracy of the SSM reduction [8]. What is commonly a drawback in conventional numerical techniques thus becomes a mechanism to enhance robustness and adaptivity.

To achieve scalability and computational efficiency, the SSM formulation was implemented using a multi-index notation [9], which significantly improves performance compared to tensor-based approaches when approximating multivariate polynomial expansions. Furthermore, adjoint-based sensitivity expressions were derived for SSMs of arbitrary order, enabling gradient-based optimization in high-dimensional design spaces, including topology optimization problems with thousands of variables [10]. This combination of analytical precision and computational tractability represents a key contribution of the proposed framework.

The capabilities of the approach were demonstrated through a range of parametric and topology optimization problems, successfully illustrating its ability to shape the nonlinear frequency-amplitude relation. Table 8.1 summarizes the computational cost associated with all the numerical examples, offering an overview of the framework's efficiency across different case studies. Collectively, these

 
M. Pozzi, F. Braghin, *Structural Optimization for Nonlinear Dynamics*, PoliMI SpringerBriefs, https://doi.org/10.1007/978-3-032-22600-6_8

**Table 8.1** Summary of the computational cost associated with the numerical examples. All times were measured on a Windows workstation equipped with an Intel Core i9-13900KS CPU @ 3.20 GHz and 128 GB of RAM @ 4200 MT/s

| Problem | Target | Problem size | Design variables |
|---|---|---|---|
| (7.2) | Backbone curve | 27 | 4 |
| (7.3) | Backbone coefficient | 366 | 10 |
| (7.5) | Frequency response | 366 | 15 |
| (7.6) | Backbone coefficient | 40198 | 40000 |
| (7.7) | Backbone coefficient | 40198 | 40000 |
| (7.8) | Backbone curve | 16359 | 8000 |
| Problem | Total time | Iterations | Iteration time |
| (7.2) | 5 s | 10 | 0.50 s |
| (7.3) | 3 min 50 s | 65 | 3.54 s |
| (7.5) | 12 min 25 s | 45 | 16.6 s |
| (7.6) | 57 min 10 s | 128 | 26.8 s |
| (7.7) | 52 min 38 s | 116 | 27.2 s |
| (7.8) | 6 h 34 min 49 s | 143 | 2 min 45.7 s |

examples show that the proposed method can serve as a versatile and reliable design tool for mechanical systems operating in strongly nonlinear regimes.

## 8.1 Outlook

Looking ahead, a natural next step involves the fabrication and experimental validation of the optimized MEMS gyroscope prototypes (Fig. 8.1). Such experiments would not only verify the predicted improvements in performance but also shed light on the practical challenges of translating computational designs into physical devices. The methodology could also be extended to more advanced MEMS architectures, such as tuning-fork or beating-heart gyroscopes [1], as well as to other classes of inertial MEMS devices, including resonators and accelerometers [4].

From a methodological standpoint, future research may explore optimization strategies targeting additional nonlinear dynamic phenomena, such as parametric resonances and mode-coupling effects, which hold potential for performance enhancement and tunable functionality [5, 7, 11]. The adjoint sensitivities derived in this work can also serve broader purposes beyond optimization, enabling systematic investigations into how parameter variations influence nonlinear response characteristics. Such analyses could pave the way for robust design and uncertainty quantification frameworks tailored to nonlinear systems [6, 12].

In the field of topology optimization, further extensions are equally promising. One direction involves developing a more efficient implementation, potentially in a compiled language such as C++, to better handle large-scale problems with higher SSM expansion orders. At present, computational times become prohibitive in

**Fig. 8.1** GDS file of the optimized gyroscope prototypes in KLayout, a 2D viewer and editor for mask layouts used in MEMS manufacturing. The left layout is the reference design, optimized without considering nonlinearities (Fig. 3.4). The middle and right layouts correspond to the drive linearization (Fig. 7.3) and mode-matching (Fig. 7.6) optimization cases.

strongly nonlinear regimes, especially when nonlinear effects concentrate in void regions. A potential remedy lies in the incorporation of regularization terms [2, 3] to stabilize low-density areas and suppress spurious nonlinear contributions. Before applying these strategies to nonlinear dynamics, however, their behavior should first be systematically studied in linear eigenfrequency problems to ensure numerical stability and reliability.

## 8.2 Final Remark

In summary, this book establishes a foundation for a systematic, computationally efficient approach to the optimization of nonlinear mechanical systems, with a particular focus on MEMS applications. By combining SSM-based model reduction, adjoint-based sensitivities, and gradient-driven optimization, the proposed framework offers a powerful tool for the analysis and design of high-performance structures operating beyond the linear regime. More broadly, it illustrates how nonlinear dynamics and optimization can be unified into a single, forward-looking paradigm for modern engineering design.

## References


1. Acar C, Shkel A (2009) MEMS vibratory gyroscopes. Springer, Berlin. https://doi.org/10.1007/978-0-387-09536-3
2. Bluhm GL, Sigmund O, Poulios K (2021) Internal contact modeling for finite strain topology optimization. Comput Mech 67:1099–1114. https://doi.org/10.1007/s00466-021-01974-x
3. Frederiksen AH, Dalklint A, Sigmund O, Poulios K (2025) Improved third medium formulation for 3d topology optimization with contact. Comput Methods Appl Mech Eng 436:117595. https://doi.org/10.1016/j.cma.2024.117595

4. Kempe V (2011) Inertial MEMS. Cambridge University Press, Cambridge. https://doi.org/10.1017/CBO9780511933899
5. Li D, Shaw SW, Polunin PM (2022) Computational modeling of nonlinear dynamics and its utility in mems gyroscopes. J Struct Dyn. 1:217–235. https://doi.org/10.25518/2684-6500.96
6. Morsy AA, Tiso P (2025) Model reduction for systems with random parameters using spectral submanifolds. Nonlinear Dyn 113:24141–24157. https://doi.org/10.1007/s11071-025-11396-6
7. Polunin PM, Shaw SW (2022) Maximizing the rate sensitivity of resonating gyroscopes using nonlinear shape optimlization. J Micromech Microeng 32:064003. https://doi.org/10.1088/1361-6439/ac6c74
8. Pozzi M, Marconi J, Jain S, Braghin F (2024) Backbone curve tailoring via lyapunov subcenter manifold optimization. Nonlinear Dyn 112:15719–15739. https://doi.org/10.1007/s11071-024-09881-5
9. Pozzi M, Marconi J, Jain S, Li M, Braghin F (2025) Adjoint sensitivities for the optimization of nonlinear structural dynamics via spectral submanifolds. Proc R Soc A Math Phys Eng Sci 481:20250244. https://doi.org/10.1098/rspa.2025.0244
10. Pozzi M, Marconi J, Jain S, Li M, Braghin F (2025) Topology optimization of nonlinear structural dynamics with invariant manifold-based reduced order models. Struct Multidiscip Optim 68:72. https://doi.org/10.1007/s00158-025-04010-1
11. Rhoads JF, Shaw SW, Turner KL (2010) Nonlinear dynamics and its applications in micro- and nanoresonators. J Dyn Syst Measur Control 132:034001. https://doi.org/10.1115/1.4001333
12. Xu Z, Kaundinya RS, Jain S, Haller G (2025) Nonlinear model reduction to random spectral submanifolds in random vibrations. J Sound Vib 600:118923. https://doi.org/10.1016/j.jsv.2024.118923

Zeitfracht Medien GmbH
Ferdinand-Jühlke-Straße 7
99095 Erfurt, Deutschland
produktsicherheit@kolibri360.de